Phase Change Materials: Design and Applications

Phase Change Materials: Design and Applications

Editor

Ioannis Kartsonakis

MDPI • Basel • Beijing • Wuhan • Barcelona • Belgrade • Manchester • Tokyo • Cluj • Tianjin

Editor
Ioannis Kartsonakis
School of Chemical
Engineering
National Technical University
of Athens
Zographos
Greece

Editorial Office
MDPI
St. Alban-Anlage 66
4052 Basel, Switzerland

This is a reprint of articles from the Special Issue published online in the open access journal *Applied Sciences* (ISSN 2076-3417) (available at: www.mdpi.com/journal/applsci/special_issues/phase_change_material_design_application).

For citation purposes, cite each article independently as indicated on the article page online and as indicated below:

LastName, A.A.; LastName, B.B.; LastName, C.C. Article Title. *Journal Name* **Year**, *Volume Number*, Page Range.

ISBN 978-3-0365-5062-6 (Hbk)
ISBN 978-3-0365-5061-9 (PDF)

Cover image courtesy of Ioannis Kartsonakis

Contents

About the Editor

Ioannis Kartsonakis

Ioannis Kartsonakis is a Chemist with a PhD in Chemistry and a MSc in Polymer science. His PhD subject was on electrochemical deposition of multifunctional conductive polymers for corrosion protection of metals (doi: 10.12681/eadd/25933), using sol-gel technique, electro-polymerization and radical polymerization. His work is focused on: Management of R&D projects; Implementation of research and development tasks in the field of corrosion and corrosion protection; Studies on physicochemical properties of nanomaterials; Synthesis and characterization of hybrid inorganic/ organic nanomaterials, coatings, fibers, cement/ polymer composites. He has experience in many European and National projects related to: Recycling of textile and plastic materials; Concrete Reinforcement; Self-healing, Antifouling, Anticorrosion, coatings and materials.

Preface to "Phase Change Materials: Design and Applications"

In recent years, science and technology have revolutionized our way of life, improving well-being and comfort for all mankind. The discovery of new materials with unique features at macro- and nano-scales has played a significant part in this advancement. The possibility of producing materials able to perform different functions and of responding to external stimuli will undoubtedly be an extremely important research area in the near future.

There is increasingly intensive research on the development of energy storage technologies due to the increased energy needs of contemporary societies. Global energy consumption, together with greenhouse gasses and CO_2 emissions, is increasing at a very fast pace due to rapid global economic growth, global population growth and the human dependence on energy-consuming appliances. This increment in global energy consumption has specific environmental implications that pose serious challenges to the environment and human health. Moreover, increased global energy consumption results in the reduction in the availability of traditional energy resources, such as coal, oil and natural gas. Therefore, there is an urgent need for the development of new systems based on the conversion and storage of sustainable and clean energy.

An effective way to increase energy systems' efficiency is through the use of latent heat thermal energy storage because of its high energy density in comparison to sensible heat storage systems. Phase change materials (PCMs) are one of the key components for the development of advanced sustainable solutions in renewable energy and engineering systems. PCMs are able to either store or release large amounts of energy, while their temperature is changed slightly or kept constant. PCMs have the ability to accumulate and store lots of energy. The high storage potential of PCMs can be realized when their phase is changed.

In order to update the field of renewable energy and engineering systems with the use of PCMs, a Special Issue entitled "Phase Change Materials: Design and Applications" is introduced. This book gathers and reviews the ten contributions (nine articles and one review), with authors from Europe, Asia and America, accepted for publication in the aforementioned Special Issue of *Applied Sciences*. This book aims to attract all researchers working in this research field, and collects new findings and recent advances in the development, synthesis, structure–activity relationships, and future applications of PCMs.

Ioannis Kartsonakis
Editor

Editorial

Special Issue on "Phase Change Materials: Design and Applications"

Ioannis A. Kartsonakis 

Laboratory of Advanced, Composite, Nanomaterials and Nanotechnology (R-Nano Lab), School of Chemical Engineering, National Technical University of Athens, Zografou Campus, 9 Heroon Polytechniou Str., 15773 Athens, Greece; ikartso@chemeng.ntua.gr; Tel.: +30-210-772-3310

Citation: Kartsonakis, I.A. Special Issue on "Phase Change Materials: Design and Applications". *Appl. Sci.* **2022**, *12*, 7770. https://doi.org/10.3390/app12157770

Received: 30 July 2022
Accepted: 1 August 2022
Published: 2 August 2022

Publisher's Note: MDPI stays neutral with regard to jurisdictional claims in published maps and institutional affiliations.

1. Introduction

In recent years, science and technology have revolutionized our way of life, improving well-being and comfort for all mankind. The discovery of new materials with unique features at macro- and nano-scales has played a significant part in this advancement. The possibility of producing materials able to perform different functions and of responding to external stimuli will undoubtedly be an extremely important research area for the foreseeable future.

There is increasingly intensive research for energy storage technologies development due to the enhanced energy needs of the contemporary societies. Global energy consumption, together with greenhouse gasses and CO_2 emissions, is increasing at a very fast pace due to rapid global economic growth, global population growth, and the human dependence on energy-consuming appliances. This increment in global energy consumption has specific environmental implications that pose serious challenges to the environment and human health. Moreover, the enhanced global energy consumption results in the reduction in the availability of traditional energy resources, such as coal, oil, and natural gas. Therefore, there is an urgent need for new systems development based on the conversion and storage of sustainable and clean energy.

An effective way for the energy systems efficiency increment is the use of latent heat thermal energy storage because of its high energy density in comparison to sensible heat storage systems. Phase change materials (PCMs) are one of the key components for the development of advanced sustainable solutions in renewable energy and engineering systems. PCMs enable either the storage or release of large amounts of energy, while their temperature is slightly changed or kept constant. PCMs have the ability to accumulate and store lots of energy. The activation of this high storage potential of PCMs is accomplished when their phase is changed.

In order to update the field of renewable energy and engineering systems with the use of PCMs, a Special Issue entitled "Phase Change Materials: Design and Applications" has been introduced. This editorial manuscript gathers and reviews the collection of ten contributions (nine articles and one review), with authors from Europe, Asia, and America accepted for publication in the aforementioned Special Issue of *Applied Sciences*.

2. Macro- and Micro-Encapsulated PCMs

The encapsulated PCMs consist of the core material, i.e., the PCM, and an organic or inorganic shell material. Depending on their size, PCMs can be characterized as nano-encapsulated when their size is less than 1 μm, micro-encapsulated when their size is between 1 and 1000 μm, and macro-encapsulated when their size is bigger than 1000 μm. Four articles have been published in this Special Issue related to the synthesis and characterization of macro- and micro-encapsulated PCMs. The micro-encapsulation of sodium nitrate ($NaNO_3$) in zinc oxide shells as PCM for high temperature latent heat storage was investigated by Neagoe et al. [1]. The aim of their work was to evaluate the potential of $NaNO_3$-ZnO microcapsules with respect to thermal energy storage applications together with cost reduction related to metal corrosion in installations and storage tanks. Charge/self-discharge experiments revealed that the $NaNO_3$ content of the microcapsules influences the amount of thermal energy stored in the corresponding interval. Moreover, it was proved that prototype-scale thermal energy storage in microcapsules can be predicted from laboratory measurements on small material samples.

In the study of Borreguero et al. [2] micro-encapsulated PCMs were fabricated based on a core of paraffin Rubitherm® RT27 and a shell that consists of ethyl–vinyl acetate and low density polyethylene. The goal of this work was the production of polyurethane elastomeric materials demonstrating thermoregulating properties due to the incorporation of the aforementioned micro-encapsulated PCMs. The obtained results revealed that the mechanical properties and the density of the synthesized elastomeric materials together with the thermal synergy effects were related to the wt% content of the micro-encapsulated PCMs in order improved structural stability and enhanced latent heat to be demonstrated.

In the work of Gschwander et al. [3] the storage capacities of PCMs emulsions based on even- and odd-numbered paraffins with different purities are investigated, taking into account the corresponding supercooling effect and phase transition behavior. The obtained results revealed that the nucleation agent and the emulsified PCM influence both the storage and supercooling factors of the PCM emulsions. It was observed that the PCM´s hydrocarbon chain and the purity of the materials affected the degree of supercooling.

3. Design and Applications

The design of the energy storage material encapsulation is one of the most important parameters that critically influence the heat transfer in charging and discharging procedures of the storage system. The study of Vérez et al. [4] examined the effect of the design of the PCM macro-encapsulation on the thermal behavior of a latent heat thermal energy storage systems during both the charging and discharging processes. The obtained results revealed that the design of the thermal energy unit should be performed and analyzed based on the requirements of the application due to the fact that the macro-encapsulation design has a relevant impact on the heat transfer.

Ohmura and his team [5] conducted experiments on the investigation of hydrate heat storage systems. An evaluation of the kinetic characteristics of thermal energy storage using tetrabutylammonium acrylate hydrate as PCMs was performed. It was found that either mechanical agitation or ultrasonic vibration improved the thermal energy storage kinetic characteristics because of the detachment of the hydrate adhesion on the heat exchanger, which could be a thermal resistance between the thermal energy storage medium and heat exchanger.

The work of Karantonis and his team [6] reports the methodology that applied the estimation of thermodynamic and kinetic information to the interaction of lytic polysaccharide monooxygenases enzymes using a filamentous fungus. This class of enzymes boost the release of oxidized products from the biomass plant. Taking into account that these enzymes are redox, their exploitation was conducted through their immobilization on electrode surfaces and investigating the parameter standard electron transfer rate constant (k_0) between the immobilized electrode and the enzyme. The obtained results revealed that it is feasible for the system to be rendered reversible and consequently to be used for bio-electrocatalytic purposes.

Additionally, Abdi and coworkers [7] numerically examined the thermal improvement of a latent thermal energy storage (LTES) component, including mini-channels as dry air passages. The thermal performance of the LTES in charging and discharging processes with ranging flow rates and a varying number of channels was estimated. The results revealed that the LTES-phase change power is enhanced by increasing the number of channels. On the other hand, a decrease in the storage capacity was observed. Moreover, it was noticed that at a constant air flow rate there is an increment of the mean heat transfer rate because of the influence of the heat transfer coefficient due to the increase in the heat transfer surface area of the increased number of channels.

A very interesting concept developed by Luna and his team [8] based on the examination of sugar alcohols as PCMs. The effect of thermal treatment on sugar alcohol erythritol thermal, chemical, and physical properties was assessed. Moreover, experiments with erythritol and its mixtures with antioxidants were performed in order to enhance its thermal stability. It was found that in the presence of air, mixtures of erythritol with antioxidant had a lower degradation rate in comparison with pure erythritol.

In another study, Nicolalde et al. [9] reported that the use of a PCM layer of 20 mm can improve the thermal comfort in the vehicle, reducing the need to use the heater or the air conditioner. It was found that the PCM reduced the temperature of the air by 9 °C when heating and by 4 °C when the temperature dropped.

Finally, in the review paper of Kartsonakis et al. [10], the principles of latent heat thermal energy storage systems with PCMs are discussed. This paper presents the research status of the materials that can be used as PCMs, together with the most effective methods for improving their thermal performance. Moreover, several passive applications in the building sector are highlighted. Finally, special attention is given to the encapsulated PCMs that consist of the core material, which is the PCM, and the shell material, which can be either organic or inorganic, and their utilization inside constructional materials.

4. Future Strategies

Although the Special Issue has been closed, more in-depth research in the field of the synthesis of PCMs and the design of corresponding components as well as their application is expected. It can be anticipated that more effective and eco-friendly applications will be demanded in large numbers in the future for the utilization of thermal energy storage. In this case, suitable strategies should be ready for consolidation and utilization.

Funding: This research received no external funding.

Acknowledgments: The Guest Editor would like to thank all the authors and peer reviewers for their fruitful and valuable contributions to this special issue. The confluence of the editorial team of *Applied Sciences* is highly appreciated.

Conflicts of Interest: The author declares no conflict of interest.

References

1. Neagoe, C.; Tudor, I.A.; Ciobota, C.F.; Bogdanescu, C.; Stanciu, P.; Zărnescu-Ivan, N.; Piticescu, R.R.; Romero-Sanchez, M.D. Demonstration of Phase Change Thermal Energy Storage in Zinc Oxide Microencapsulated Sodium Nitrate. *Appl. Sci.* **2021**, *11*, 6234. [CrossRef]
2. Borreguero, A.M.; Izarra, I.; Garrido, I.; Trzebiatowska, P.J.; Datta, J.; Serrano, Á.; Rodríguez, J.F.; Carmona, M. Thermal and Mechanical Behavior of Elastomers Incorporated with Thermoregulating Microcapsules. *Appl. Sci.* **2021**, *11*, 5370. [CrossRef]
3. Gschwander, S.; Niedermaier, S.; Gamisch, S.; Kick, M.; Klünder, F.; Haussmann, T. Storage Capacity in Dependency of Supercooling and Cycle Stability of Different PCM Emulsions. *Appl. Sci.* **2021**, *11*, 3612. [CrossRef]
4. Vérez, D.; Borri, E.; Crespo, A.; Mselle, B.D.; de Gracia, Á.; Zsembinszki, G.; Cabeza, L.F. Experimental Study on Two PCM Macro-Encapsulation Designs in a Thermal Energy Storage Tank. *Appl. Sci.* **2021**, *11*, 6171. [CrossRef]
5. Kiyokawa, H.; Tokutomi, H.; Ishida, S.; Nishi, H.; Ohmura, R. Thermal Energy Storage Performance of Tetrabutylammonium Acrylate Hydrate as Phase Change Materials. *Appl. Sci.* **2021**, *11*, 4848. [CrossRef]
6. Zouraris, D.; Karnaouri, A.; Xydou, R.; Topakas, E.; Karantonis, A. Exploitation of Enzymes for the Production of Biofuels: Electrochemical Determination of Kinetic Parameters of LPMOs. *Appl. Sci.* **2021**, *11*, 4715. [CrossRef]
7. Abdi, A.; Chiu, J.N.; Martin, V. Numerical Investigation of Latent Thermal Storage in a Compact Heat Exchanger Using Mini-Channels. *Appl. Sci.* **2021**, *11*, 5985. [CrossRef]
8. Alferez Luna, M.P.; Neumann, H.; Gschwander, S. Stability Study of Erythritol as Phase Change Material for Medium Temperature Thermal Applications. *Appl. Sci.* **2021**, *11*, 5448. [CrossRef]
9. Nicolalde, J.F.; Cabrera, M.; Martínez-Gómez, J.; Salazar, R.B.; Reyes, E. Selection of a PCM for a Vehicle's Rooftop by Multicriteria Decision Methods and Simulation. *Appl. Sci.* **2021**, *11*, 6359. [CrossRef]
10. Podara, C.V.; Kartsonakis, I.A.; Charitidis, C.A. Towards Phase Change Materials for Thermal Energy Storage: Classification, Improvements and Applications in the Building Sector. *Appl. Sci.* **2021**, *11*, 1490. [CrossRef]

applied sciences

Article

Demonstration of Phase Change Thermal Energy Storage in Zinc Oxide Microencapsulated Sodium Nitrate

Ciprian Neagoe [1,*], Ioan Albert Tudor [1], Cristina Florentina Ciobota [1], Cristian Bogdanescu [1], Paul Stanciu [1], Nicoleta Zărnescu-Ivan [1], Radu Robert Piticescu [1,*] and Maria Dolores Romero-Sanchez [1,2]

[1] National R&D Institute for Nonferrous and Rare Metals–IMNR, 077145 Pantelimon, Romania; atudor@imnr.ro (I.A.T.); crusti@imnr.ro (C.F.C.); cbogdanescu@imnr.ro (C.B.); pstanciu@imnr.ro (P.S.); nicoleta.zarnescu@imnr.ro (N.Z.-I.); md.romero@applynano.com (M.D.R.-S.)
[2] Applynano Solutions, S.L., Parque Científico de Alicante, 03690 Alicante, Spain
* Correspondence: ciprian.neagoe@imnr.ro (C.N.); rpiticescu@imnr.ro (R.R.P.)

Featured Application: Microencapsulation of alkali metal nitrates for latent heat storage in concentrated solar power plants has the potential to reduce salt corrosion in storage tanks and installations.

Abstract: Microencapsulation of sodium nitrate ($NaNO_3$) as phase change material for high temperature thermal energy storage aims to reduce costs related to metal corrosion in storage tanks. The goal of this work was to test in a prototype thermal energy storage tank (16.7 L internal volume) the thermal properties of $NaNO_3$ microencapsulated in zinc oxide shells, and estimate the potential of $NaNO_3$–ZnO microcapsules for thermal storage applications. A fast and scalable microencapsulation procedure was developed, a flow calorimetry method was adapted, and a template document created to perform tank thermal transfer simulation by the finite element method (FEM) was set in Microsoft Excel. Differential scanning calorimetry (DSC) and transient plane source (TPS) methods were used to measure, in small samples, the temperature dependency of melting/solidification heat, specific heat, and thermal conductivity of the $NaNO_3$–ZnO microcapsules. Scanning electron microscopy (SEM) and chemical analysis demonstrated the stability of microcapsules over multiple tank charge–discharge cycles. The energy stored as latent heat is available for a temperature interval from 303 to 285 °C, corresponding to onset–offset for $NaNO_3$ solidification. Charge–self-discharge experiments on the pilot tank showed that the amount of thermal energy stored in this interval largely corresponds to the $NaNO_3$ content of the microcapsules; the high temperature energy density of microcapsules is estimated in the range from 145 to 179 MJ/m^3. Comparison between real tank experiments and FEM simulations demonstrated that DSC and TPS laboratory measurements on microcapsule thermal properties may reliably be used to design applications for thermal energy storage.

Keywords: phase change materials; sodium nitrate; thermal conductivity; microencapsulation; latent heat; thermal energy storage

Citation: Neagoe, C.; Tudor, I.A.; Ciobota, C.F.; Bogdanescu, C.; Stanciu, P.; Zărnescu-Ivan, N.; Piticescu, R.R.; Romero-Sanchez, M.D. Demonstration of Phase Change Thermal Energy Storage in Zinc Oxide Microencapsulated Sodium Nitrate. *Appl. Sci.* **2021**, *11*, 6234. https://doi.org/10.3390/app11136234

Academic Editor: Ioannis Kartsonakis

Received: 19 May 2021
Accepted: 24 June 2021
Published: 5 July 2021

Publisher's Note: MDPI stays neutral with regard to jurisdictional claims in published maps and institutional affiliations.

1. Introduction

Thermal energy storage is important for the development of industrial and domestic applications that optimize energy production and use. Efficiency of thermal energy storage systems depends, among other factors, on the properties of the storage material to accumulate and release energy by either changing its temperature, in sensitive heat storage, or undergoing a phase transition, in latent heat storage (for a review, see Sarbu and Sebarchievici, 2018) [1]. Storing energy as latent heat may provide a large amount of energy at almost constant temperature, an important feature in some technical applications that require close control over temperature, and may be a robust and inexpensive alternative to temperature control automation by electronic devices.

Selection of a material suitable for latent heat energy storage (see Podara et al., 2021) [2] considers multiple criteria such as temperature for phase change, latent heat, thermal stability of the material, containment, and cost (Hussam et al., 2020) [3]. Sodium and potassium nitrate salts as phase change materials (PCM) cover the range of relatively high temperatures of interest (300–400 °C) for concentrated solar plants (CSP). Sodium nitrate undergoes a solid–liquid phase change transition at 307 °C (Xu et al., 2018) [4], whereas potassium nitrate melts at higher temperature, 336 °C (Mohammad 2017; D'Aguanno et al., 2018) [5,6]. The latent heats of solid–liquid transition in sodium and potassium nitrate salts are 5.9 and 11.8 times, respectively, smaller than the latent heat for LiF solidification (1.04 MJ/kg at 849 °C). The tradeoff between energy storage capacity and the melting temperature of the material may consider that (a) higher or lower temperatures may not be technically suited for a specific application, (b) temperature increase is associated with accelerated corrosion of containers and pipes, and (c) the heat loss rate from the storage container depends directly on the temperature gradient across the container shell.

For sodium nitrate PCM, limitations of tank charge–discharge rates are related to the low thermal conductivity of the storage material (Bellan et al., 2015) [7]. Micro and macroencapsulation may increase the thermal conductivity of the storage material if a highly conductive shell material is used (Huang et al., 2019; Cáceres et al., 2017) [8,9]. Similarly, composite materials are explored, where a matrix of chemically stable, thermally conductive material mitigates the problem of a highly corrosive and low conductive inorganic salt (or salt mixture) used to store the thermal energy (Courbon et al., 2020; Podara et al., 2021) [2,10].

The thermal properties of the storage material may be combined with appropriate design of the storage tank using in silico experiments of computational fluid dynamics (CFD) (Al-abidi et al., 2013) [11]. The finite element method (FEM) with common software packages (Comsol Multiphysics) was used to compare various inorganic salt PCMs and different suitable PCM-encapsulation materials to render economically viable solutions that would meet the requirements of thermal energy storage (TES) for CSP (Cáceres et al., 2017) [9].

Micro and nanoencapsulation has been frequently reported for low temperature regime PCMs, mainly using organic polymers as shell materials. For high temperature applications (e.g., CSP), manufacturing micrometer-scale shells that are stable over many PCM melting–solidification cycles is challenging for at least two reasons: (a) coatings made by the versatile organic polymer technology do not withstand high temperatures; and (b) water-based chemistry to prepare shells may dissolve the PCM core, which is typically an inorganic salt/salt mixture [12]. SiO_2 coating of micrometer-size alkaline nitrate cores was recently demonstrated by several groups (Zhang et al., 2018; Romero-Sanchez et al., 2018; Lee and Jo, 2020) [12–14]. A method for microencapsulation of nitrate salts in zinc oxide shells was developed within a project funded by the National Authority for Scientific Research in Romania (project acronym ENERHIGH), as described in our previous papers (Romero-Sanchez et al., 2019; Tudor et al., 2018) [15,16]. Briefly, in that work, KNO_3–ZnO microcapsules were prepared from raw materials (zinc nitrate tetrahydrate and potassium nitrate) by the solvothermal method. Spherical morphology of the ZnO-covered KNO_3 microparticles was observed after spray drying, with particle sizes of only a few microns.

The aim of this work is to evaluate the capacity of $NaNO_3$–ZnO microcapsules to store thermal energy and the possibility to retrieve it at high temperatures. To improve the scalability of microcapsule fabrication, a novel protocol for microencapsulation was elaborated and tested, consisting in direct ZnO coating of $NaNO_3$ particles using commercially available $NaNO_3$ grains (relatively large, up to 0.5 mm size) and ZnO powder (0.1 to 2 μm particles) as raw materials. Up to 22.1 kg of $NaNO_3$–ZnO microcapsules were thus prepared. Results from experiments in a 16.7 L cylinder-shaped prototype tank are presented, which show that microcapsules store significant amounts of thermal energy in solid–liquid $NaNO_3$ transitions. To facilitate further development of novel PCM materials for energy storage, a platform for combined in silico prototype-scale investigation was built. For a simplified tank FEM model, thermal transfer is presented in simulations of tank discharge and compared with real temperature data from prototype tank experiments on microcapsules. The goal is to understand and predict material performance in thermal storage applications, starting from material characterization by laboratory methods.

2. Materials and Methods

Chemicals. Analytical grade reagents were used in all experiments. Sodium nitrate was purchased from Roth, Germany. Zinc oxide was provided by Chimreactiv, Romania.

Microcapsules. $NaNO_3$ was encapsulated in ZnO shells at 4:1 mass ratio, using a simplified protocol derived from our previous work (Tudor et al., 2018) [16]. Basically, commercially available $NaNO_3$ powder grains, ZnO, and ethanol were mixed and stirred to achieve homogenous dispersion of components. The mixture was then subjected to a thermal treatment for 1 h at 200 °C in atmospheric air. Microcapsules were characterized by scanning electron microscopy and energy dispersive X-ray spectroscopy to determine their structure and demonstrate encapsulation.

Chemical analysis of the thermal energy storage material was determined by flame atomic absorption spectrometry (ISO 11047:1999) for Na and Zn and wet chemistry (ISO 5664: 2001) for NO_3^-. Samples were taken from the tank at 14 different positions in the storage material.

Scanning Electron Microscopy (SEM) and Energy Dispersive X-ray Spectroscopy (EDS) were performed on microcapsules using an SEM–EDS Quanta 250 scanning electron microscope (FEI Company, Eindhoven, The Netherlands) equipped with the Element Silicon Drift Detector Fixed and Element EDS Analysis Software Suite APEX™ 1.0, EDAX, Mahwah, NJ, USA. Thus, by EDS, the atomic composition of a selected spot was determined. In the report from the instrument, the abbreviations C K, O K, ZnL, NK, NaK, and AlK are used to indicate that the data are based on detection of energies for specific orbital transitions, corresponding to elements: C, O, Zn, N, Na, and Al, respectively. The elemental composition of each spot is reported as relative weight (%), atom concentration (%), or the net intensity (a.u.) of the respective element-specific peak. For microcapsules, the carbon % reported in analysis is background from the sample immobilization device, and was subtracted from further calculations.

Thermal analysis. Thermal conductivity and thermal diffusivity/specific volumetric heat determinations were made by the transient plane source method (TPS-2200 from Hot Disk, Sweden, model 5082 mica sensor). In laboratory determinations, microcapsule powder was poured to fill a 200 mL stainless steel beaker, the sensor was embedded in the middle, the assembly was placed in the empty (air-filled) storage tank, then exposed to multiple temperature cycles in the 25 to 350 °C temperature interval. The thermal conductivity and specific volumetric heat curves were recorded only after the second cycle of microcapsule melting–solidification, when the microcapsule material was a compact porous material block; thus, the thermal parameters are representative for the material formed after multiple energy storage cycles. Temperature monitoring was achieved by tank thermocouples and corrected to report in-place temperatures using calibrated readings for the sensor resistivity.

Differential Scanning Calorimetry (DSC) was performed using two instruments: SET-SYS Evolution from Setaram and DSC F3 Maia from Netzsch, Germany. Microgram amounts of $NaNO_3$ and microcapsules were heated–cooled in cycles to determine melting and crystallization temperatures and enthalpies.

Prototype tank. The thermal storage tank is a right circular cylinder (referred to as the tank cylinder), 0.55 m length, 0.2 m diameter, made of 2 mm-thick stainless steel (SS 316), accommodating temperature sensors in inbuilt sheaths protruding transversally up to the longitudinal axis. A thermal conductivity sensor is located in a holder at the center-bottom of the tank. The cylinder is wrapped in 0.2 m thick Ravaber mineral foam insulation. A stainless-steel envelope covers the insulation to a final cylinder length of 0.95 m and diameter of 0.6 m. Material loading gates (three 20 mm diameter holes) are placed at the top of the storage tank and accessible from outside by cap closed tubes that span the thermal insulation. The heat exchanger is made of a SS 316 tube (12 mm outer diameter, 10 mm inner diameter, 5.04 m length), shaped as a helical-coil wrapped around the longitudinal axis in 15 loops (120 mm diameter). Mineral oil is used as a heat transfer fluid (HTF) at a temperature that is maintained by an external recirculating pump with heating/cooling capability. Temperatures were continuously monitored (digital recording) by eight type J thermocouples (T1–T8) placed as indicated in Figure 1. For the 50 to 400 °C measurement interval, the thermocouples were linearly calibrated to work within a ±1.5 °C reading error. T3, T4, and T5 measured temperatures inside the tank, at the central longitudinal axis, and they protruded to their measurement spot through tank-fixed one-end-closed tubular sheets (5 mm outer diameter, 3 mm inner diameter, 0.5 m long). The relative positions of the protruding thermocouple sheets and heating coil loops were designed to maximize the separation space and avoid contact. Temperatures were monitored across the energy storage material for two sections that are transversal to the tank cylinder, by placing T1 and T2 directly on the outer surface of the tank cylinder, at the same longitudinal positions as their corresponding T4–T3 (respectively) pair. T7 and T8 were in direct contact with the heating coil at the tank-envelope entry and exit points. The installation is pictured in a previous publication (Tudor et al., 2019) [17].

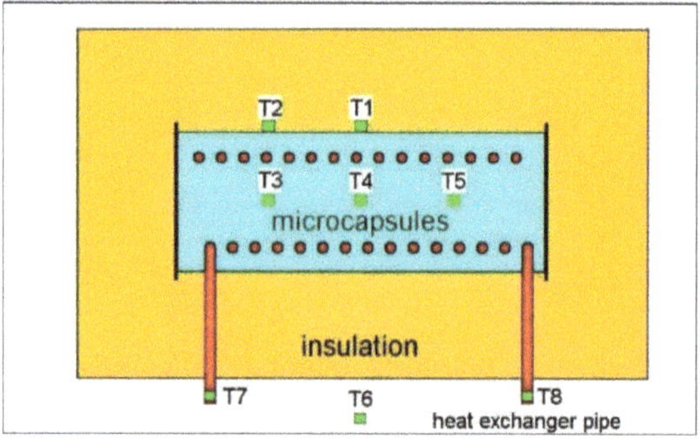

Figure 1. Schematic drawing of the prototype tank (median longitudinal section—viewed from the top): light blue—microcapsules, green—T1 to T8 thermocouples, yellow—insulation, red—heat exchanger. T6 is the room temperature thermocouple.

Tank heating–cooling cycles. Precise amounts of microcapsules were successively loaded into the storage tank (as powder) to total 15.75, 19.232, and 22.067 kg. After each load, the tank was subjected to several charge–self-discharge cycles. A cycle consists of a heating phase, when HTF (360–375 °C) circulates inside the heat exchanger until T1 to T5 readings are all above the melting temperature of sodium nitrate, then it follows the self-discharging phase when the HTF pump is off and the tank is left to cool by itself downward to room temperature.

Adapted calorimetric method. To assess the amount of thermal energy available as latent heat in the storage tank, a heat flow calorimetry method was developed. It is based on temperature monitoring in key points inside and outside the tank, thus estimating the heat loss through the thermal insulation of the tank.

The method considers a quasi-static approximation to Fourier's law:

$$Q/dt = -K\Delta T_{\text{in-out}},$$

with dt as the time interval for measurements, short enough to approximate a constant $\Delta T_{\text{in-out}}$. The approximation is intended to simplify the calculation, as it was determined that the use of differential form to account for $\Delta T_{\text{in-out}}$ changes does not significantly improve the precision for a narrow temperature interval around 300 °C. In the tank setup, at the time-point t, $\Delta T_{\text{in-out}} = tT1 - tT6$, where tT1 and tT6 are the temperatures indicated by T1 and T6, respectively.

For two different material loads of the tank with masses m_1 and m_2, it may be written:

$$Q_{m1}/dt_1 = -K\,\Delta T_{\text{in-out}} \tag{1}$$

$$Q_{m2}/dt_2 = -K\,\Delta T_{\text{in-out}} \tag{2}$$

where K is a proportionality coefficient for heat transfer from the storage tank to the environment. A third equation considers the caloric capacities of the tank components for the two loading m_1 and m_2, resulting in:

$$Q_{m2}\text{-}Q_{m1} = (m_2 - m_1)\,c_p\,dT \tag{3}$$

where dT is a small temperature change that occurred in dt_1, the same that occurred in dt_2 (temperature decay in experimental T1, immediately after the 8–9 h plateau interval—characteristic for latent storage); and c_p is the specific heat of the storage material (microcapsules), determined by DSC just below the solidification offset temperature. For tank calculations, dT is monitored from the T1 curve; the difference between T1 and T2 affects $\Delta T_{\text{in-out}}$ estimation less than 8% in the relevant temperature interval. In the tank experiments, a T1 curve just below solidification (point indicated by the offset temperature in T4) is almost a straight line, therefore the dt (respectively, dT) can be quite large without affecting the integration of heat flows to Q_{m1} and Q_{m2} heats for Equation (3).

Thermal analysis using transient nonlinear FEM.

Model and implementation. The tank model was implemented in Microsoft Excel spreadsheets, with the FEM simulation step progressing on the columns of a simulation table. Visual representation (time course, temperature dependency, etc.) of input parameters and output results was available in charts. To estimate the heat transfer, the tank was modeled as a cylinder, then FEM was applied for 10 concentrical layers (cells), imposing T1 boundary conditions from the experimental dataset. Only conductive transfer was taken into consideration, as suggested by Guo and Zhang (2008) [18] for sodium nitrate. Heat transfer during PCM solidification was calculated by the apparent capacity method, derived from the enthalpy method (Kousksou et al., 2012; Khattari et al., 2020) [19,20]. Thus, microparticle heat transfer during solidification was modeled based on conversion of the solidification heat into specific heat. Thermal conductivity and specific heats were then approximated on specific intervals by polynomial functions on temperature to best fit the experimental data. For a narrow T4 interval at the onset of solidification, a constant thermal conductivity value, 2× higher than measured, was in place for T4 simulations, as it was the single major contributor to simulation–measurement overlap on that interval, probably due to some convective heat transfer by microcapsule deformation in the tank.

Simulations start from the presumed steady-state equilibrium reached at the end of the charging phase. Thus, inside the tank, the radial symmetric temperature distribution was set to range linearly from maximal temperature (T4) to the lowest temperature (T1). After few cycles of initial calculation, the temperature differences between consecutive cycles converged in each calculation point, therefore reaching a more realistic radial temperature distribution than initially assumed. This calculated quasi-steady state was attained with time–space resolution of the FEM mesh (time step 0.25 s and layer thickness 10–2 m) that ensured agreement with the second principle of thermodynamics.

Model limitations and corrections.

Symmetry assumptions. The radial symmetry of the tank is broken on partial tank filling because of the flat horizontal lining at the top of the storage material. In our model, this affects the cross-layer transfer in the material, introducing parallel heat transfer (microcapsule–air in parallel with microcapsule–microcapsule) for a same FEM layer. To overcome this without increasing FEM mesh complexity, for adjacent layers we calculated the effective transfer area (representing only the layers contact within the microcapsule-material block), and use it to express the exchanged heat. Thus, radial heat transfer between consecutive layers occurred on local temperature difference and effective transfer area. Transversal heat transfer occurred for each layer across the insulation, also considering the effective transfer area and T2–T6 temperature gradient (same for all layers, which is justified by a conductive steel plate that contacted all layers at the side-end). Symmetry across a vertical transversal plane intersecting the tank center was assumed. At each simulation step, the transferred thermal energy was assigned to respective layers, and then a new set of temperatures was calculated.

Experimental temperature readings were shifted from real values by thermocouple shielding. This occurred in the one-end-closed sheath tubes (T3, T4, T5) used to protrude with thermocouples at deep places inside the storage material. Therefore, with the tank empty, we performed a verification experiment for thermocouple readings calibration. A thermocouple was placed at T3 through the protruding tube and another thermocouple, in the same spot, from the inside the tank. The sheath of T3 separated the two thermocouples. The tank was then heated and T3 inside–outside values were recorded. A lag in T3 outside readings was observed. Temperatures were not corrected, as the value is change-rate dependent; instead this temperature difference was acknowledged in the calculations (by the adapted calorimetric method) for the stored energy, and reported distinctively for maximally corrected and uncorrected readings.

Input and output data for simulation: A cycle of FEM simulation consisted in complete calculation of the cell array. For each cell, the cycle output was (a) the amount of energy transferred between the respective cell and the surrounding cells, (b) the end-cycle temperature, and (c) the mass ratio between melt/solid PCM in a cell. Microcapsule experiments (Hot Disk and DSC) provided the thermal conductivity, specific heat, and heat of fusion of the storage material at various temperatures. In FEM, feed-in data for those temperature-dependent thermal parameters were generated by polynomial interpolation (up to grade 6) on multiple segments in the 30 to 380 °C interval. The heat of fusion was expressed as apparent specific heat, to account for the temperature - dependent heat-flow profile in the narrow temperature interval from solidification onset to solidification offset. Simulated T4 temperature evolution was compared with the corresponding experimental values.

3. Results

3.1. Microencapsulation of NaNO$_3$ Is Persistent in the Storage Tank, over Multiple Thermal Cycles

Our simplified protocol for NaNO$_3$ microencapsulation produced microcapsules 100 to 300 microns in diameter. The microcapsule shell is made of ZnO particles arranged in a nonuniform layer ranging from a few nanometers up to 3 microns thickness. Based on relevant SEM samples, a gross estimation of the whole population of microparticles suggests a majority of well-coated and thermally stable microparticles, with some prone to collapse from thermal cycles. To test their collective thermal stability, the NaNO$_3$–ZnO microcapsules were loaded in the storage tank and exposed to multiple melting–solidification cycles.

Scanning electron microscopy (Figure 2) was employed to demonstrate that microencapsulation was effective and that it is lasting after charge–discharge cycles in the tank. Just after preparation, the microcapsule material was a powder consisting of relatively large (~300 μm) grains. Some of the grains presented a nonuniform gray surface, with light and dark spots (Figure 2A). Figure 2D confirmed that light gray areas are ZnO particles encapsulating NaNO$_3$ grains (dark gray). The correlation was established by EDS (Figure 2C) and grayscale analysis on the small surface spots was pictured in Figure 2E. Thus, elemental concentration (%) of each spot was plotted against the spot intensity on the digital image (grayscale: 0 to 255). On microcapsules that are similar to the microcapsule in the center of Figure 2A, due to the interruptions in the ZnO cover layer, the thickness of the layer can be estimated at the edges of the 2D grain shape. Figure 2B shows the microcapsule-based material after multiple thermal storage cycles in the tank. It compacted into a porous block with microcapsules (smaller than 100 μm) that are often distinctly arising and are bonded by interstitial NaNO$_3$. Whereas before storage cycles some microcapsules were nonuniformly covered with ZnO particles, with areas of exposed NaNO$_3$ core, partially covered microcapsules are absent from material samples after repeated melting–solidification.

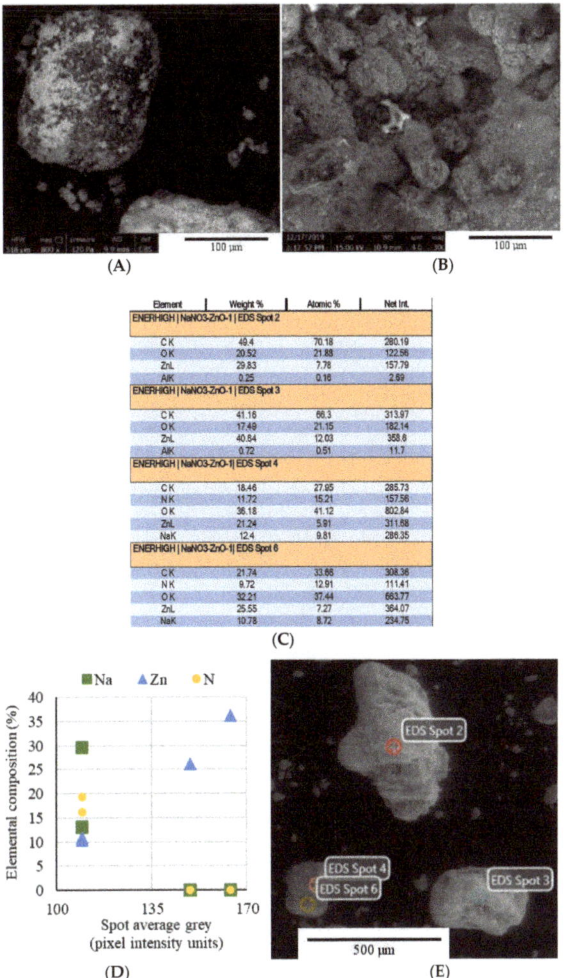

Element	Weight %	Atomic %	Net Int.
ENERHIGH | NaNO3-ZnO-1 | EDS Spot 2			
C K	49.4	70.18	280.19
O K	20.52	21.88	122.56
ZnL	29.83	7.78	157.79
AlK	0.25	0.16	2.69
ENERHIGH | NaNO3-ZnO-1 | EDS Spot 3			
C K	41.16	66.3	313.97
O K	17.49	21.15	182.14
ZnL	40.84	12.03	358.8
AlK	0.72	0.51	11.7
ENERHIGH | NaNO3-ZnO-1| EDS Spot 4			
C K	18.46	27.95	285.73
N K	11.72	15.21	157.56
O K	36.18	41.12	802.84
ZnL	21.24	5.91	311.88
NaK	12.4	9.81	286.35
ENERHIGH | NaNO3-ZnO-1| EDS Spot 6			
C K	21.74	33.66	308.36
N K	9.72	12.91	111.41
O K	32.21	37.44	663.77
ZnL	25.55	7.27	364.07
NaK	10.78	8.72	234.75

Figure 2. SEM micrograph of $NaNO_3$–ZnO microcapsule before (**A**) and after repeated melting–solidification cycles (**B**). EDS analysis (**C**) of 4 spots in a same picture (**E**), confirming (**D**) that light grey spots contain Zn and do not contain Na or N.

The chemical composition was analyzed on small samples taken from 14 spots at various depths (0 to 45 mm from the front section surface) and radial positions within the storage material. The $NaNO_3$–ZnO mass ratio diminished in most of the spots from an initial value 4:1 to 3.4:1 ($\pm$0.14 standard deviation). Table 1 groups in three intervals the $NaNO_3$:ZnO mass ratio detected in the 14 spots. Most of the microcapsule samples had a mass ratio around 4:1, showing that the chemical composition was preserved locally. For the spots outside the average range, a plausible explanation is that during the liquid phase of the storage cycles, the $NaNO_3$ content of weakly formed microcapsules leaked in the interstitial space and bonded at solidification with stable microcapsules, in a porous material. The ZnO coating of the collapsing microcapsules redistributed to surroundings. Gravitational stratification occurred, proven by composition analysis of six spots from the upper side versus six spots from the lower side of the tank. Thus, average $NaNO_3$–ZnO mass ratios of 3.4 $\pm$ 0.1 at the upper half of the tank and 4.9 $\pm$ 1.6 at the bottom were obtained.

Table 1. Analysis of microparticle composition stability by NaNO$_3$–ZnO mass ratio after multiple phase-change thermal cycles.

NaNO$_3$:ZnO (Mass Ratio)	3.4:1 ÷ 4:1	4:1 ÷ 8:1	above 60:1
Count (no. of spots)	9	3	2

3.2. Temperature-Dependent Thermal Properties of NaNO$_3$–ZnO Microcapsules

Latent heats of NaNO$_3$ solid–liquid transitions were determined by DSC experiments in microcapsules and pure salt (Figure 3). For microcapsules, the melting heat was 102.564 kJ/kg and the solidification heat was 94.5 kJ/kg. In addition, the specific heat at 284.8 °C was 0.824 kJ/kg. Pure NaNO$_3$ required 124.5 kJ/kg for solidification. For microcapsules, the temperature intervals for melting and solidification are [t_{onset}, t_{offset}] = [303, 319] °C and [302.5, 285] °C, respectively (Figure 3A). This does not significantly differ from pure salt, (Figure 3B—our data). Thus, in the DSC experiments at 10 °C/min cooling rate, supercooling tends to be less pronounced in microcapsules, and clearly reduced compared with similar experiments on pure NaNO$_3$, as reported by Mohammad in his Ph.D. thesis, Table 7.2.2.1. [5]. The higher solidification heat (modulus) reported there (167 kJ/kg) for NaNO$_3$ is partly related to the fact that our measurements were done for repeated melting–solidification cycles, with latent heat in the first-cycle consistently higher than the rest; converging to the reported value). The onset and offset temperatures (solidification) were used in simulations and in calorimetric calculations. In FEM simulations, the normalized DSC curve profile and solidification latent heat were principal determinants of the T4 temperature evolution, whereas calorimetric calculations used onset and offset temperatures from DSC to precisely identify the interval for latent heat contribution to the experimental T4 curve.

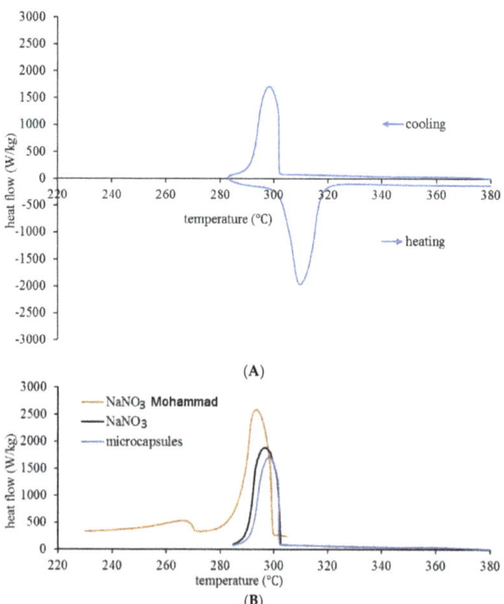

Figure 3. (**A**) DSC heating - cooling curves for NaNO$_3$–ZnO microcapsules. (**B**) Comparison of DSC solidification curves (10 °C /min cooling rate) for pure NaNO$_3$ (Mohammad [5] and own data) and microcapsules.

Figure 4 presents the specific heat and the thermal conductivity of microcapsules in comparison with pure NaNO$_3$ (own experiment and from Bauer et al., 2013 [21]). In the temperature interval from 303 to 285 °C the Hot Disk values are not shown (data from DSC were used in calculations). In the temperature interval 30–270 °C, the thermal conductivity of the microcapsules is around 0.3 W/m·K. At 280 °C, a minor decrease in thermal conductivity was noticed, probably associated with phase transition for $NaNO_3$ crystalline structure. Above 300 °C, a sudden increase in the thermal conductivity was measured, correlating with PCM melting. The room temperature density of the microcapsule-based storage material after several melting–solidification cycles was calculated, considering the density decrease when increasing temperature, as reported for $NaNO_3$ [21]. A value of 1.8 t/m^3 was inferred and used for FEM simulations (at temperatures just below the liquid–solid transition temperature).

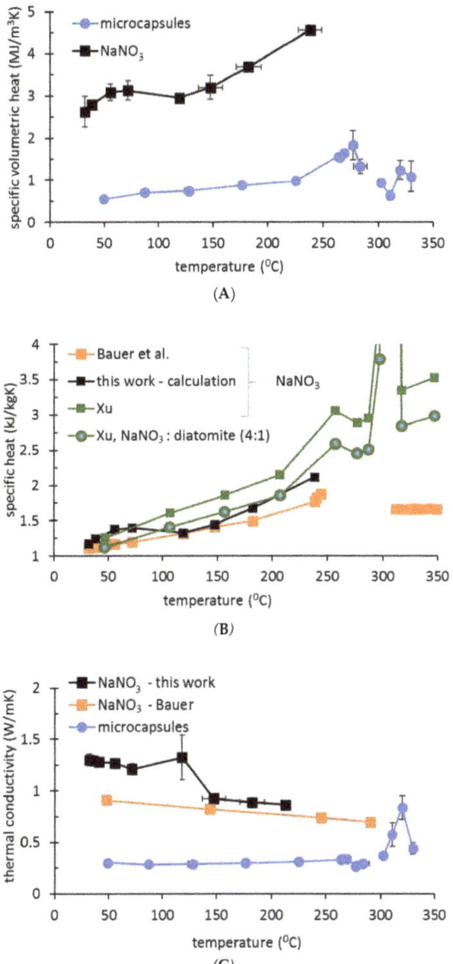

Figure 4. Thermal properties of $NaNO_3$–ZnO microcapsules in comparison with pure $NaNO_3$. Hot Disk determinations (own data) of volumetric specific heat (**A**) and thermal conductivity (**C**), in comparison with other works. (**B**) Specific heat of pure sodium nitrate and of a composite mixture of $NaNO_3$: diatomite (same weight ratio as microcapsules). External data are from Bauer et al., 2013 [21] and Xu et al., 2017 [4].

3.3. Availability of High-Temperature Thermal Energy from Microcapsule Storage

In a typical charge–self-discharge experiment (Figure 5) the highest temperatures are almost always recorded by T4, in the center of the tank, near the heating coil. T1 is placed on the surface of the stainless-steel shell, on the same transversal section as T4, and therefore used in conjunction with T4 to monitor the temperature in radial direction across the storage material. The presumed left–right symmetry from the center of the tank is consistent with T3 versus T5 readings.

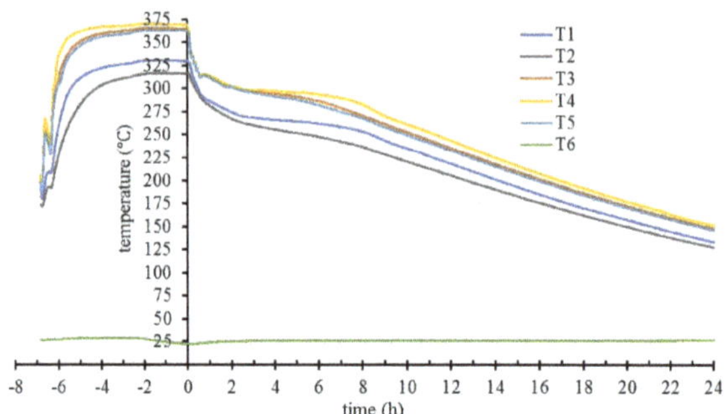

Figure 5. T1 to T6 temperatures readings (positions in the prototype tank indicated in Figure 1) during a heating–cooling cycle.

In Figure 6, plateaus of relatively constant T4 and T1 temperatures are noted shortly after the beginning of the free-cooling phase of cycles. For T4, this corresponds to the liquid to solid phase transition when latent energy is released from microparticle cores at rates that dominate over sensitive heat. After PCM solidification phase, the temperature decay in tank is controlled by two variables: the caloric capacity of the tank and the heat transfer through tank insulation. For a narrow temperature interval, using linear fit slopes of T1 curves corresponding to 19.23 and 22.07 kg (taken near the PCM offset temperature), K value was determined. Based on the T1 plateau, by extrapolating a constant K towards t = 0 h, the heat transfer rate was calculated, thus estimating the latent PCM energy stored in the tank.

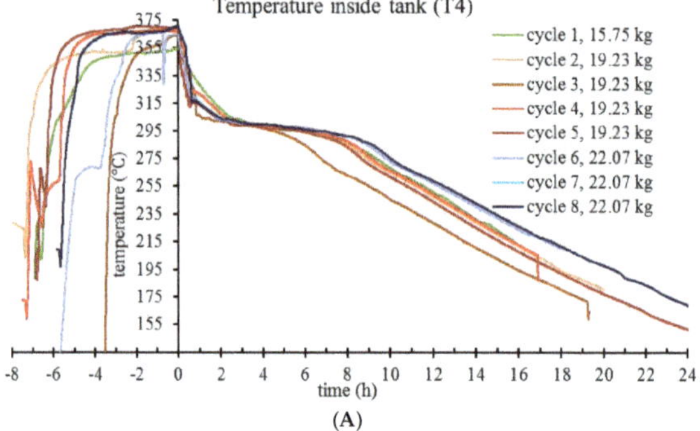

Figure 6. *Cont.*

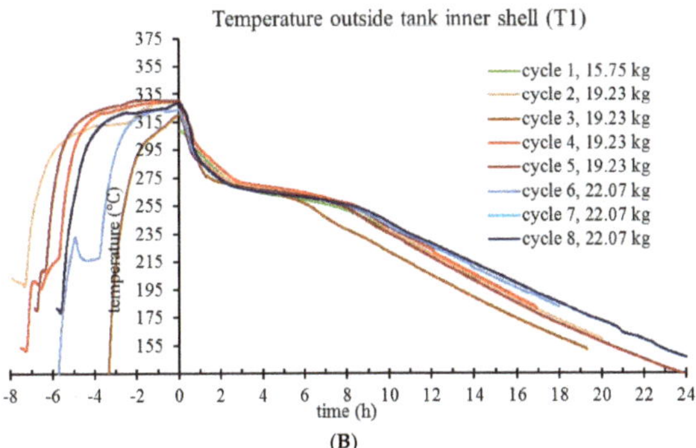

Figure 6. T4 (**A**) and T1 (**B**) temperature evolution during charge–self-discharge cycles of the prototype tank, for progressively increased loads of storage material.

In the evaluation of the thermal energy storage capacity of the tank, two loads of microcapsules were considered: 19.23 and 22.07 kg. T4 monitors the solidification in the innermost spot of the tank, therefore informative for the end of the latent heat release phase. At high temperature change rates, T4 readings are affected by nonequilibrium, with measured temperature differences up to 15 °C for the interval of interest. Thus, to protrude with the T3–T5 thermocouples at their deep spots into the storage material, the tank has three thin tubes welded on the longitudinal side. The tubes are closed at the protruding end and they prevent direct contact between thermocouples and microcapsules. In an experiment to determine the magnitude of correction, another thermocouple was inserted into the air-filled storage tank, with the tip measuring the temperature in the same spot as T4, but from inside the tank. In the experiment presented in Figure 7, for $T_{real} = 275$ °C, the $T_{measured}$ is 260 °C - showing up to 15 °C temperature differences. This is explained by nonequilibrium measurements for T3 - T5 during tank heating. Correction of T4 on this basis is not straightforward, since at plateau temperature (in the solidification phase) the temperature equilibrium between material and sensor is attained. Therefore, a correction was not applied to thermocouple readings. Instead, in Figure 8, when reporting the energy available as latent heat (by adapted calorimetric method) in tank discharge, a 15 °C uncertainty on T4 readings was explicitly acknowledged for the solidification offset temperature, and the corresponding energy values given as a maximum estimate.

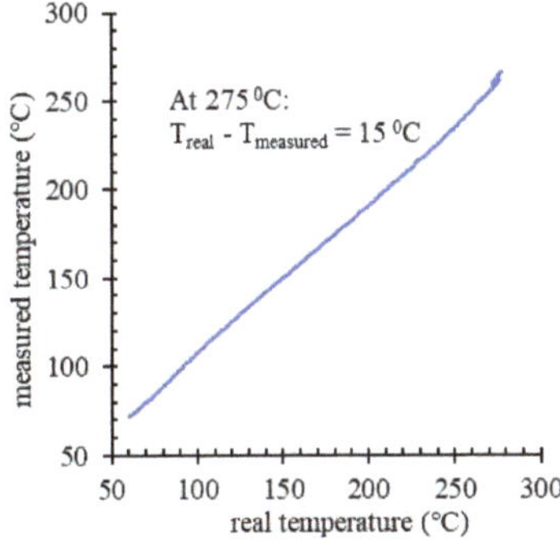

Figure 7. The accuracy of T3–T5 thermocouple response to temperature dynamics inside the storage tank.

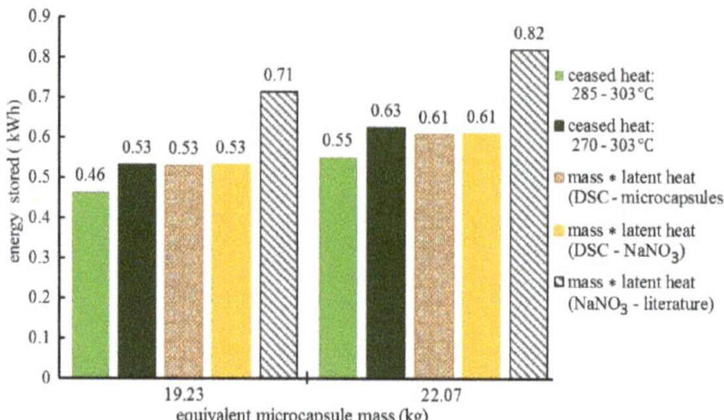

Figure 8. Storage tank evaluation of the latent heat thermal energy storage in microcapsules by adapted calorimetric method (ceased heat) and latent heat measurements at microscale.

Based on the latent heats, a summary is provided in Figure 8 for the estimates of the energy stored as latent heat in the partially filled storage tank. Results are presented for microcapsules in comparison with pure $NaNO_3$ at equal PCM content mass. Two tank loads were considered: 19.23 and 22.07 kg of microcapsules, with a net $NaNO_3$ content of 15.38 and 17.65 kg, respectively. In the calorimetric method adapted, the phase-change-stored thermal energy is the heat ceased by the tank in the temperature interval for solidification, with T4 as the monitor. Solidification in microcapsules was considered for uncorrected [285, 303] °C and maximally corrected [270, 303] °C temperature intervals. A specific heat of 0.824 kJ/ kg resulted from DSC measurements and it was used to implement Equation (3). The first two columns at each tank load represent the estimations of the stored energy by this adapted calorimetric method. The ceased heat is compared with tank-level calculations based on (a) microscale DSC measurements of the latent (solidification) heat for microcapsules and $NaNO_3$, and (b) the reference (literature) latent heat (167 kJ/kg) for $NaNO_3$.

The columns labeled as mass * latent heat represent estimates of the tank-stored PCM solidification energy, and are provided by the mass–latent heat product for microcapsules (DSC) and pure $NaNO_3$ (our DSC and literature). The mass of pure $NaNO_3$ corresponds to the PCM content of microcapsules.

3.4. Tank Discharge Insight View by FEM Simulation

In our FEM model tank, for 19.23 and 23.06 kg microcapsules loads, the discharge phase (of the experimental charge–self-discharge cycles) was simulated. T1 and T2 readings from real experiments were the boundary temperatures for tank inner shell: T1 on longitudinal and T2 on transversal margins. The thermal properties of the storage material were kept identical for the two load conditions, with respect to their temperature dependency. The thermal energy transferred at each simulation step was calculated, resulting in the temperature evolution for each layer. T4 experimental readings were compared with corresponding temperatures in simulations. Experimental T1 was imposed in simulations as the temperature of the outer (tenth) layer to predict temperatures at each of the nine inner concentric layers. T2–T6 was used for all layers as the temperature gradient on the lateral side of the tank. In Figure 9, a good fit between experimental and simulated T4 curves indicate that the model and parameters are accurate, thus validating thermal conductivities, specific heats, and solidification heat data. The main parameter controlling the overlap between simulated and experimental T4 in the plateau area was the latent heat of liquid–solid transition. The specific heat of microcapsules provided the slope of the T4 curve below solidification offset temperature. At the beginning of the discharge phase, the fast T4 decay could only be reproduced in simulations with thermal conductivity >200% of the Hot Disk value. Various factors contribute to this unfitted section of the T4 decay, among which may be considered T4 thermocouple shielding and convective heat transfer via gravitational leakage of low viscosity interstitial (nonencapsulated) PCM.

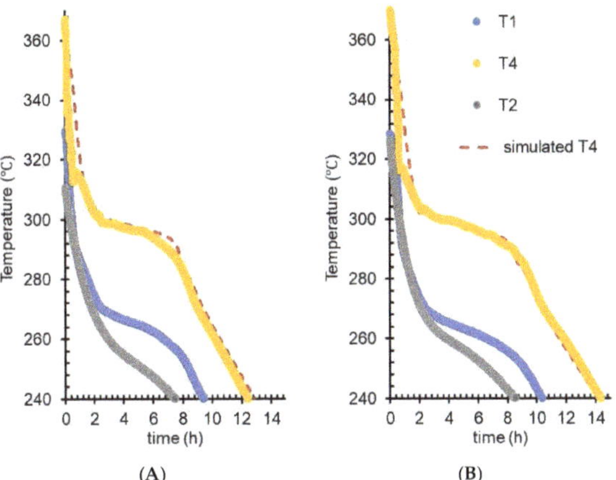

Figure 9. FEM simulations of T4 (red segmented line) in the self-discharge experiment, compared with experimental data (filled circles) for T1 (blue), T2 (gray), and T4 (yellow): (**A**) 19.23 kg microcapsules load and (**B**): 22.07 kg load.

4. Discussion

PCM microencapsulation aims to enhance the properties of materials for thermal energy storage and holds potential to provide effective solutions for applications in heating systems (Huang 2019) [8]. Microencapsulation of $NaNO_3$ in ZnO shell for high temperature thermal energy storage may reduce the investment cost of using high-grade stainless steel (Fernandez et al., 2014) [22] to contain the highly corrosive salt (Tudor et al., 2018) [16]. In Tudor et al., 2018 [16], a solvothermal process was used to produce nanometer-thick zinc oxide shells around potassium nitrate cores, obtaining microcapsules of few microns size. In the present paper, preparation of larger size (100 to 300 μm) microparticles is reported, each encapsulated in a layer of ZnO nanoparticles. Due to changes in the microencapsulation protocol (solvothermal synthesis was eliminated), the ZnO envelopes reported here typically show highly variable thickness, ranging from few nanometers to few microns.

Initially uneven, the shell ZnO layer of microcapsules seems to uniformly distribute upon melting–solidification, and thus resists multiple charging–discharging cycles. In this view of a potential for improvements in ZnO deposition, a 4:1 $NaNO_3$: ZnO mass ratio may be lower than optimal, where optimization means that for complete covering of the microparticle, the ratio should maximize the PCM mass on the expense of shell thickness. Despite this suboptimal PCM:shell ratio, our novel simplified protocol for microencapsulation may minimize the initial investments in specialized equipment for manufacturing. The observed thermal stability of microcapsules reported here may be explained by the fact that melting–solidification of PCM is able to select, after few cycles, structurally stable microcapsules. With the leaking PCM, compaction of the storage material at macroscale may occur, increasing the thermal energy density in storage cycles.

The temperature dependency for density, thermal conductivity, and specific heat of pure $NaNO_3$ is available from multiple sources (Xu et al., 2018; D'Aguanno et al., 2018; Bauer et al., 2017) [4,6,21]. Therefore, it was possible to compare our data (Figure 4B) in the same graph with results reported by Xu et al., 2018 [4] for $NaNO_3$ and for a mixture of diatomite: $NaNO_3$ at the same weight ratio as our $NaNO_3$:ZnO (4:1).

From Figure 4A, which shows the volumetric specific heat, it is possible to calculate the specific heat (gravimetric) of the material by considering the temperature dependency of the material density. Since the density for the microcapsule-based material was not measured for high temperatures, a temperature variation similar to that described for $NaNO_3$ [21] was assumed. In the temperature interval from 50 to 284 °C, a value between 300 and 1020 J/kg·K resulted from the calculations of the specific heat of microcapsules; particularly important at 284 °C, where the specific heat for microcapsules is 801 J/kg·K. FEM simulations (Figure 9) validated 750 J/kg·K as the only value of the specific heat that allows consistent overlap below the solidification T_{offset} with the experimental T4 slope (for both tank loads). For Equation (3) in the adapted calorimetric method, the 824 J/kg·K value (measured by DSC) was used; this is near 801 J/kg·K value deduced from volumetric specific heat, and close to the 750 J/kg·K value from FEM validation. Thus, FEM simulations not only reproduce the experimental temperatures, but are useful to estimate the accuracy of other methods (e.g., DSC and adapted calorimetric method) by selecting, from multiple fitting (simulation) sessions, a specific set of values for the material thermal parameters.

Multiple applications are designed to simulate thermal transfer (Al-abidi et al., 2013) [11]. While designing a standalone CFD application [17] capable of reproducing complex geometries, it was observed that a simplified cylinder model of the storage tank is sufficiently informative and fast to produce end-simulation results. The FEM simulations require as input the (a) tank geometry, (b) thermal properties of the storage material (from laboratory measurements), and (c) time course of temperature on the tank shell (monitored in tank charge–discharge experiments). The simulation output is a time course for temperatures and thermal energy distribution in the virtual tank, confirmed by thermocouple temperature monitoring in the center of the prototype storage tank.

The FEM model allows to evaluate to which degree accuracy is needed for thermal property inputs (thermal conductivity and specific heat), and whereas polynomial fits are good descriptors for experimental data. The unusual choice of Microsoft Excel for the implementation of FEM simulations is based on the cylindrical symmetry and the simplicity of the storage tank, allowing for massive reduction in the mesh size. The validity of the simplification was tested by the convergence of the result, in each cell, for an interval of time steps ranging from 0.1 to 0.5 s. Although in terms of computing resources needed at a glance, Excel simulation approach is more demanding than an equivalent runtime simulation application, it has the advantage of facilitating development of features such as graphically augmented, transparent, and explicative input and output data (Tomimura et al., 2009) [23].

The main question to which this study responded is if high temperature thermal energy may effectively be recovered from storage in microcapsules. In a simplistic evaluation of the tank-stored latent energy, the latent heat of microcapsules is multiplied by their mass, providing a target value (maximal) for the potential of the tank. In real thermal storage cycles, a PCM material is not uniformly exposed to energy transfer during tank charging; melting of all available PCM mass depends on factors such as thermal conductivity of the microcapsules, charging time, and temperature in the heat exchanger circuit. This is evident in the charge–self-discharge cycle 3 (Figure 6), when a plateau temperature (quasi-steady state) was not reached at charging time, therefore the high-temperature discharge phase was shorter. Ideally, when charging is complete and all $NaNO_3$ content is liquid, most of its latent heat is retrievable at temperatures that correspond to latent heat storage.

Figure 8 presents the high temperature energy that can be retrieved from $NaNO_3$–ZnO microcapsules, which corresponds to specific energy storage of 80.5 kJ/kg as the low estimate and 99.5 kJ/kg as the high estimate. These values are comparable with specific energies for other inorganic salt-based PCMs [24,25]. The corresponding energy density range (1.45×105 to 1.79×105 kJ/m^3), considering 1800 kg/m^3 for the microcapsule material density (after melting–solidification cycles), is also as expected.

Noticeable differences of the $NaNO_3$ tank-stored energy exist (Figure 8) when calculations are based on literature data versus this work's DSC measurements. This is explained by a drop of the $NaNO_3$ latent heat after the first melting–cooling cycle, probably due to partial nitrate decomposition (Bauer et. al 2013) [21].

Estimations of the specific energy incorporate several sources of error: (a) the lag in T3–T5 temperature readings presented in Figure 7, (b) small (unintended) variations of experimental conditions in the tank setup (even small sudden changes in the ambient light were spotted in temperature traces), and (c) the precision of the thermocouples that were linearly calibrated for a large temperature interval. Instead of presenting data means from multiple cycles (with error bars), the graphs in Figure 6 show temperature evolutions for each cycle, which is more informative for tendencies that fall under statistical thresholds. With four cycles on 19.23 kg load and three cycles on 22.07 kg load, data show no pattern for temperature–trace variation between cycles of a same load, whereas temperature traces corresponding to different loads are clustered. By reporting a range for the estimated specific energy density of microcapsules (from Figure 8 data) instead of a data mean and error bars, this work acknowledges limitations in the precision of the adapted calorimetric method to measure the latent heat that is stored in the prototype tank. Nevertheless, alternative methods such as heat flow balance focused on monitoring the difference between T8 and T7 during active HTF pumping/extraction are seriously affected by the thermocouple precision ($\pm 1.5\ ^\circ$C) and the possibility to monitor HTF flow rate, while still dependent on estimation of heat flow through tank insulation.

The workflow for design research and development of a new solution for latent heat energy storage was charted by Whiffen and Riffat (2013) in their Figure 5 [24]. It comprises several parallel activities already reported for the ENERHIGH project [13,15–17], such as selection of materials in the appropriate range, design, and development of the heat exchanger and construction of the prototype. Here, laboratory measurements of microcapsule thermophysical properties and prototype experimentation converge to FEM simulation, in a platform for integrated material design and testing, to be used for enhancement of relevant properties and optimization of material use in product applications.

To better evidence the process and optimize charging/extraction for intermittent energy sources, FEM simulations with adequate geometry details can produce time-resolved 3D maps of heat transfer and the possibility to explore various real-use scenarios. Work is in progress to design and implement a software application that enables FEM simulations for complex geometries of the storage tank and spatial resolutions of only a few millimeters (Tudor et al., 2019) [17]. Implementing methodological solutions from the current study, a simulation application may be able to reproduce thermal transfers with higher fidelity during melting–solidification. The nonlinear temperature dependencies of thermodynamic parameters (thermal conductivity and apparent specific heat) may be obtained from laboratory determinations and used as input in FEM simulations to produce overlap with the experimental temperature curves; the outcome is exemplified in Figure 9.

5. Conclusions

This work presents laboratory and prototype-scale characterization of an innovative PCM for thermal energy storage: (i) thermal properties of $NaNO_3$–ZnO microcapsules; (ii) a prototype tank demonstration of phase-change thermal energy storage; and (iii) a methodological framework for design, scaled manufacturing, and prototype testing of microencapsulated $NaNO_3$. Methodological novelties are: (a) effective manufacturing of microcapsules, (b) a method to measure the high temperature thermal energy stored in a prototype tank, and (c) a software tool (as a document template) to analyze and predict the dynamics of PCM energy storage during charge–discharge cycles of a cylindrical storage tank.

In summary:

- $NaNO_3$–ZnO microcapsules were prepared by a simple and easy to scale protocol, then characterized for structural and thermal properties;
- Charge–discharge experiments of the storage tank demonstrated the availability of high-temperature thermal energy stored as latent heat in $NaNO_3$–ZnO microcapsules;
- The temperature interval to access this energy in the pilot tank is 303 to 285 $^\circ$C, as predicted by DSC experiments at microscale;

- By the adapted calorimetric method, we estimated that in-tank high temperature specific energy of microcapsules is in the range 80.5–99.5 kJ/kg, in line with expectations for inorganic salts;
- Thermal transfer simulations using the Microsoft Excel template setup for transient nonlinear FEM confirmed that prototype-scale thermal energy storage in microcapsules can be predicted from laboratory measurements on small material samples.

Future work may use the simulation–prototyping platform to develop microencapsulated phase change materials with enhanced thermal conductivity, ensuring optimal heat transfer rates for specific applications.

Author Contributions: Conceptualization, R.R.P., M.D.R.-S., C.N., and P.S.; methodology, C.N., M.D.R.-S., R.R.P., and P.S.; investigation, C.N., I.A.T., C.B., C.F.C., P.S., and N.Z.-I.; writing—original draft preparation, C.N.; writing—review and editing, C.N., R.R.P., and M.D.R.-S.; supervision, C.B., R.R.P. and M.D.R.-S.; project administration, R.R.P. and M.D.R.-S.; funding acquisition, R.R.P. and M.D.R.-S. All authors have read and agreed to the published version of the manuscript.

Funding: The research results are based on the financial support from National Authority for Scientific Research in Romania: ENERHIGH project (IDP_37_776, MySMIS104730, Contract 93/09.09.2016 under the Competitive Operational Program 2014–2020) and PN 19190101/2019–2022.

Acknowledgments: Authors are grateful to Adrian Mihail Motoc for scientific advice and administrative support. Excellent technical assistance was provided by Eng. Maria Magdalena Stoiciu.

References

1. Sarbu, I.; Sebarchievici, C. A Comprehensive Review of Thermal Energy Storage. *Sustainability* **2018**, *10*, 191. [CrossRef]
2. Podara, C.V.; Kartsonakis, I.A.; Charitidis, C.A. Towards Phase Change Materials for Thermal Energy Storage: Classification, Improvements and Applications in the Building Sector. *Appl. Sci.* **2021**, *11*, 1490. [CrossRef]
3. Jouhara, H.; Żabnieńska-Góra, A.; Khordehgah, N.; Ahmad, D.; Lipinski, T. Latent thermal energy storage technologies and applications: A review. *Int. J. Thermofluids* **2020**, *5*, 100039. [CrossRef]
4. Xu, G.; Leng, G.; Yang, C.; Qin, Y.; Wu, Y.; Chen, H.; Cong, L.; Ding, Y. Sodium nitrate—Diatomite composite materials for thermal energy storage. *Sol. Energy* **2017**, *146*, 494–502. [CrossRef]
5. Mohammad, M.B. High Temperature Properties of Molten Nitrate Salts for Solar Thermal Energy Storage Application. Ph.D. Thesis, Swinburne University of Technology, Melbourne, Australia, 2016. Available online: http://hdl.handle.net/1959.3/425444 (accessed on 30 June 2021).
6. D'Aguanno, B.; Karthik, M.; Grace, A.N.; Floris, A. Thermostatic properties of nitrate molten salts and their solar and eutectic mixtures. *Sci. Rep.* **2018**, *8*, 1–15. [CrossRef] [PubMed]
7. Bellan, S.; Alam, T.E.; Gonzalez-Aguilar, J.; Romero, M.; Rahman, M.M.; Goswami, D.Y.; Stefanakos, E.K. Numerical and experimental studies on heat transfer characteristics of thermal energy storage system packed with molten salt PCM capsules. *Appl. Therm. Eng.* **2015**, *90*, 970–979. [CrossRef]
8. Huang, X.; Zhu, C.; Lin, Y.; Fang, G. Thermal properties and applications of microencapsulated PCM for thermal energy storage: A review. *Appl. Therm. Eng.* **2019**, *147*, 841–855. [CrossRef]
9. Cáceres, G.; Fullenkamp, K.; Montané, M.; Naplocha, K.; Dmitruk, A. Encapsulated Nitrates Phase Change Material Selection for Use as Thermal Storage and Heat Transfer Materials at High Temperature in Concentrated Solar Power Plants. *Energies* **2017**, *10*, 1318. [CrossRef]
10. Courbon, E.; D'Ans, P.; Skrylnyk, O.; Frère, M. New prominent lithium bromide-based composites for thermal energy storage. *J. Energy Storage* **2020**, *32*, 101699. [CrossRef]
11. Al-Abidi, A.A.; Bin Mat, S.; Sopian, K.; Sulaiman, M.; Mohammad, A.T. CFD applications for latent heat thermal energy storage: A review. *Renew. Sustain. Energy Rev.* **2013**, *20*, 353–363. [CrossRef]
12. Zhang, H.; Balram, A.; Tiznobaik, H.; Shin, D.; Santhanagopalan, S. Microencapsulation of molten salt in stable silica shell via a water-limited sol-gel process for high temperature thermal energy storage. *Appl. Therm. Eng.* **2018**, *136*, 268–274. [CrossRef]
13. Romero-Sánchez, M.D.; Piticescu, R.-R.; Motoc, A.M.; Arán-Ais, F.; Tudor, A.I. Green chemistry solutions for sol–gel microencapsulation of phase change materials for high-temperature thermal energy storage. *Manuf. Rev.* **2018**, *5*, 8. [CrossRef]
14. Lee, J.; Jo, B. Surfactant-free microencapsulation of sodium nitrate for high temperature thermal energy storage. *Mater. Lett.* **2020**, *268*, 127576. [CrossRef]

15. Romero-Sanchez, M.D.; Piticescu, R.R.; Motoc, A.M.; Popescu, M.; Tudor, A.I. Preparation of microencapsulated KNO_3 by solvothermal technology for thermal energy storage. *J. Therm. Anal. Calorim.* **2019**, *138*, 1979–1986. [CrossRef]
16. Tudor, A.I.; Motoc, A.M.; Ciobota, C.F.; Ciobota, D.N.; Piticescu, R.R.; Romero-Sánchez, M.D. Solvothermal method as a green chemistry solution for micro-encapsulation of phase change materials for high temperature thermal energy storage. *Manuf. Rev.* **2018**, *5*, 4. [CrossRef]
17. Tudor, A.I.; Neagoe, C.; Piticescu, R.-R.; Romero-Sanchez, M.D. Microencapsulated PCMs for thermal energy storage in the range 300–500 °C: Pilot-testing. *IOP Conf. Ser. Mater. Sci. Eng.* **2019**, *572*, 012066. [CrossRef]
18. Guo, C.; Zhang, W. Numerical simulation and parametric study on new type of high temperature latent heat thermal energy storage system. *Energy Convers. Manag.* **2008**, *49*, 919–927. [CrossRef]
19. Kousksou, T.; Jamil, A.; El Omari, K.; Zeraouli, Y.; Le Guer, Y. Effect of heating rate and sample geometry on the apparent specific heat capacity: DSC applications. *Thermochim. Acta* **2011**, *519*, 59–64. [CrossRef]
20. Khattari, Y.; El Rhafiki, T.; Choab, N.; Kousksou, T.; Alaphilippe, M.; Zeraouli, Y. Apparent heat capacity method to investigate heat transfer in a composite phase change material. *J. Energy Storage* **2020**, *28*, 101239. [CrossRef]
21. Bauer, T.; Pfleger, N.; Laing, D.; Steinmann, W.-D.; Eck, M.; Kaesche, S. 20-High-Temperature Molten Salts for Solar Power Application. In *Molten Salts Chemistry: From Lab To Applications*; Lantelme, F., Groult, H., Eds.; Elsevier: Oxford, UK, 2013; pp. 415–438. [CrossRef]
22. Fernández, A.G.; Galleguillos, H.; Pérez, F.J. Thermal influence in corrosion properties of Chilean solar nitrates. *Sol. Energy* **2014**, *109*, 125–134. [CrossRef]
23. Tomimura, T.; Koito, Y.; Torii, S.; Ishizuka, M. Application of Excel to Thermal Analysis of Electronic Equipment: Program-Less Analysis and Visualization of Its Process by Using Spreadsheet of Excel. In Proceedings of the ASME 2009 InterPACK Conference, San Francisco, CA, USA, 19–23 July 2009; Volume 2, pp. 527–533. [CrossRef]
24. Whiffen, T.R.; Riffat, S.B. A review of PCM technology for thermal energy storage in the built environment: Part I. *Int. J. Low Carbon Technol.* **2012**, *8*, 147–158. [CrossRef]
25. Dincer, I.; Rosen, M.A. Energy Storage Systems. In *Thermal Energy Storage Systems and Applications*, 2nd ed.; Wiley & Sons: Chichester, UK, 2011; pp. 57–92.

Article

Thermal and Mechanical Behavior of Elastomers Incorporated with Thermoregulating Microcapsules

Ana M. Borreguero [1], Irene Izarra [1], Ignacio Garrido [2], Patrycja J. Trzebiatowska [3], Janusz Datta [4], Ángel Serrano [5], Juan F. Rodríguez [1] and Manuel Carmona [1,*]

1   Department of Chemical Engineering, University of Castilla—La Mancha, Av. Camilo José Cela s/n, 13004 Ciudad Real, Spain; Anamaria.Borreguero@uclm.es (A.M.B.); ireneizarraperezuclm@gmail.com (I.I.); Juan.RRomero@uclm.es (J.F.R.)
2   Department of Applied Mechanics and Engineering Projects, University of Castilla—La Mancha, Av. Carlos III s/n, 45071 Toledo, Spain; Ignacio.Garrido@uclm.es
3   Department of Environmental Technology, Faculty of Chemistry, University of Gdansk, Wita Stwosza 63, 80-952 Gdansk, Poland; patrycja.jutrzenka-trzebiatowska@ug.edu.pl
4   Department of Polymer Technology, Chemical Faculty, Gdansk University of Technology, G. Narutowicza Street 11/12, 80-233 Gdansk, Poland; janusz.datta@pg.edu.pl
5   Centro de Investigación Cooperativa de Energías Alternativas (CIC energiGUNE), Basque Research and Technology Alliance (BRTA), Parque Tecnológico de Alava, Albert Einstein 48, 01510 Vitoria-Gasteiz, Spain; aserrano@cicenergigune.com
*   Correspondence: Manuel.CFranco@uclm.es; Tel.: +34-926-295-300

Abstract: Polyurethane (PU) is one of the principal polymers in the global plastic market thanks to its versatility and continuous improvement. In this work, PU elastomeric materials having thermoregulating properties through the incorporation of microcapsules (mSD-(LDPE·EVA-RT27)) from low-density polyethylene and vinyl acetate containing paraffin®RT27 as PCM were produced. Elastomers were synthesized while varying the molar ratio [NCO]/[OH] between 1.05 and 1.1 and the microcapsule (MC) content from 0.0 to 20.0 wt.%. The successful synthesis of the PUs was confirmed by IR analyses. All the synthesized elastomers presented a structure formed by a net of spherical microparticles and with a minimum particle size for those with 10 wt.% MC. The density and tensile strength decreased with the MC content, probably due to worse distribution into the matrix. Elastomer E-1.05 exhibited better structural and stability properties for MC contents up to 15 wt.%, whereas E-1.1, containing 20 wt.% MC, revealed mechanical and thermal synergy effects, demonstrating good structural stability and the largest latent heat. Hence, elastomers having a large latent heat (8.7 J/g) can be produced by using a molar ratio [NCO]/[OH] of 1.1 and containing 20 wt.% mSD-(LDPE·EVA-RT27).

Keywords: polyurethane elastomers; microencapsulated PCMs; thermal properties; mechanical properties

Citation: Borreguero, A.M.; Izarra, I.; Garrido, I.; Trzebiatowska, P.J.; Datta, J.; Serrano, Á.; Rodríguez, J.F.; Carmona, M. Thermal and Mechanical Behavior of Elastomers Incorporated with Thermoregulating Microcapsules. *Appl. Sci.* **2021**, *11*, 5370. https://doi.org/10.3390/app11125370

Academic Editor: Ioannis Kartsonakis

Received: 17 May 2021
Accepted: 7 June 2021
Published: 9 June 2021

1. Introduction

Polyurethane (PU) stands out as one of the most demanded polymers, and an exponential growth of its market is expected, which could reach USD 105.2 billion by 2025 [1]. This growth is based on the constant development of enhanced PUs by the improvement of their properties for covering new applications. PU properties can be modified by varying the amount and type of polyols and isocyanates and the rest of the additives employed in their synthesis, which also allows for classifying them into two main groups: foams (flexible or rigid) and coatings, adhesives, sealants and elastomers (CASEs) [2]. Among PUs, elastomers can be also classified as cast elastomers, thermoplastic polyurethane elastomers, millable polyurethane gums and microcellular elastomers [3].

Thermoplastic polyurethane elastomers are one of the PUs with the fastest growth, since they are extremely adaptable to a high number of applications such as sports footwear, athletics tracks, auto body side molding, automotive lumbar supports and electronic products [3,4]. In most of these applications, the mechanical and thermal properties are essential to achieve proper human comfort and electronic device performance and durability. As for the rest of the PUs, the enhancement of the elastomers' quality and characteristics is related to the amount and type of their main constituents and additives [5]. The main constituents are polyester or polyether polyols (soft segments) and a diisocyanate whose chain is extended with a low molecular weight diol (hard segment) [5].

Regarding the additives, there has been an increased tendency to synthesize and characterize various clay-containing polymer nanocomposites (CPNs) for improved CASE applications for the last 10–15 years [6,7]. By focusing on elastomer polymers improved by additives, Mondal et al. [6] reported the employment of poly(ethylene-co-octene)-poly(ethylene-co-vinyl acetate) as a hybrid polymer additive for poly(ethylene-co-octene)-elastomers (POEs). The addition of only 0.5 wt.% of this additive into the POE improved the tensile strength, modulus, elongation at break, stress relaxation and hysteresis with respect to the virgin POE. Furthermore, they added organically modified montmorillonite (OMt) to the POE (0.5–1.0 wt.%), resulting in a remarkable improvement in appearance compared with the virgin POE.

The addition of cellulose nanocrystals (CNCs) into a styrene–butadiene rubber (SBR) elastomer was assayed by Annamalai et al. [8], improving the tensile storage moduli (E'). Banerjee et al. studied the addition of $CaSO_4$ as filler into an epoxidized natural rubber (ENR) matrix, observing a significant improvement in the mechanical properties, reaching a 100% increase in the dynamic modulus and the thermal stability.

Han et al. [9] introduced a novel reactive rubber composite made by compounding magnesium oxide (MgO) powder with hydrogenated nitrile butadiene rubber (HNBR) using different MgO concentrations of 0, 14, 28 and 40% v/v. The elastic modulus value tripled, reaching 80 MPa, while doubling in volume for the rubber filled with 40% MgO by volume.

Regarding the thermal properties, the enrichment of the thermoregulating capacity of elastomers to enhance the thermal comfort of the end users is garnering important attention. The thermoregulating capacity of PU foams applied in the footwear and mattress industries or automobile and construction sectors has been improved by the incorporation of microcapsules containing phase change materials (PCMs) [10,11]. PCMs can absorb or release the energy equivalent with their latent heat when the temperature undergoes or overpasses their melting points. Most PCM applications require their previous encapsulation or PCM stabilization before the inclusion in the final system to avoid leakage during the melting process [12]. In the case of microcapsules, their shell type, size and amount of used microencapsulated PCMs determine their final distribution into the polyurethane matrix [13].

Recently, some authors have incorporated PCMs into elastomers. The most relevant examples found are related to the addition of PCMs into the polymeric matrix. Armstrong et al. [14] introduced form-stable PCMs from fatty acids in elastomer copolyesters called phase-change elastomer gels (PCEGs). They explored the thermal characteristics of PCEG films, wherein the copolyester grade, gel composition and the fatty acid concentration from 30 to 70 wt.% obtained latent heats within 10–100 J/g and without hysteretic thermal cycling.

On the other hand, Juarez et al. [15] added microencapsulated PCMs (melting temperature of 52 °C) using concentrations from 1 to 10 wt.% on an elastomeric matrix of styrene–ethylene/butylene–styrene (SEBS) material with a remarkable effect on the thermal regulation of SEBS while keeping good resistant and ductile properties for PCM concentrations of 1–5 wt.%.

Chriaa et al. [16] investigated the SEBS elastomer using shape-stabilized phase change material by absorbing hexadecane into the network of the SEBS and coating it with a low-density polyethylene (LDPE). They studied four mass fractions of hexadecane/SEBS (80/5, 75/10, 65/20, 55/30, w/w %), employing 15% LDPE with respect to the melt-mixing method. They found that the thermal properties of the elastomers improved with the PCM content, varying from 106.15 kJ/kg to 179.76 kJ/kg for the composite containing 80% PCM and maintaining good properties despite the huge PCM content.

Considering the feasibility of LDPE for producing elastomer materials, in this work, thermoregulating microcapsules containing the paraffin wax Rubitherm® RT27 with a shell from LDPE and ethyl-vinyl acetate (EVA) with an average particle size of 10 μm (mSD-(LDPE·EVA-RT27)) were selected [13]. On the other hand, for first time, a polyurethane elastomeric matrix was chosen for the microencapsulated PCMs' incorporation, since it is one of the most versatile polymers. mSD-(LDPE·EVA-RT27) contents from 0.0 to 20.0 wt.% were incorporated, also varying the main constituent proportion, isocyanate and polyol for [NCO]/[OH] molar ratio values of 1.05 and 1.1. The effect of these variables on the structural, mechanical and thermal properties was evaluated, looking for the elastomer with the best thermal energy storage capacity that maintained good values for the mechanical properties.

2. Materials and Methods

Poly(tetramethylene ether)glycol (PTMG 2000) from Sigma-Aldrich was used as the polyol. Diphenylmethane-4,4′-diisocyanate (MDI) was purchased from Interchemol, and 1,4-diazabicyclo[2.2.2]octane (DABCO) from Sigma-Aldrich and 1,4-butanediol (BD) from Brenntag were used as the catalyst and chain extender, respectively. Spherical thermoregulating microcapsules containing Rubitherm® RT27 with a shell from LDPE and EVA produced by the spray drying technique (mSD-(LDPE·EVA-RT27)) with an average particle size of 10 mm and a latent heat of 86.47 J/g were used as fillers.

2.1. Elastomer Synthesis

Elastomeric composites were manufactured by using two different molar ratios between the isocyanate and hydroxyl groups ([NCO]/[OH]) with values of 1.05 and 1.1. The synthesis was realized by a two-step method. Before being used, the PTMG was vacuum dried for 1.5 h at 90 °C. In the first step, a prepolymer was produced through the reaction between PTMG and MDI (molar ratio polyol:diisocyanate = 1:4). The reaction was carried out at 85 °C for 1.5 h. In the second step, the prepolymer chains were extended by using the corresponding 1,4-butanediol to ensure the desired [NCO]/[OH] molar ratio with 0.3 wt.% 1,4-diazabicyclo[2.2.2]octane and five different microcapsule values (0.0, 5.0, 10.0, 15.0 and 20.0 wt.%). Finally, the prepared elastomers and their composites were cured at 100 °C for 24 h.

2.2. Characterization Techniques

2.2.1. Fourier Transform Infrared Spectroscopy (FTIR)

The chemical structures of the elastomers were studied by using a Varian 640-FT-IR spectrometer in the range from 600 to 4000 cm^{-1}, at a spectral resolution of 8 cm^{-1} and with 16 scans.

2.2.2. Scanning Electron Microscopy (SEM)

The elastomers' morphologies and filler distributions were depicted by means of scanning electron microscopy (SEM) with a FEI QUANTA 250, working between 5 and 30 kV.

2.2.3. Density and Porosity

The density and porosity of the elastomers were determined with a helium pycnometer (EI Accupyc II 1334) by the gas displacement method and according to ASTM D6226.

2.2.4. Thermogravimetric Analyses (TGAs)

The thermal stability of the paraffin, mSD-(LDPE·EVA-RT27) and the composites were evaluated by thermogravimetric analysis with TA Instruments equipment model SDT Q600. The used conditions for the analyses were a heating rate of 10 °C/min from room temperature to 700 °C under a nitrogen atmosphere.

2.2.5. Differential Scanning Calorimetry (DSC)

This technique allowed us to obtain the elastomers' latent heat and the homogeneity of the mSD-(LDPE·EVA-RT27) distribution by taking three samples from different elastomers zones. DSC analyses were performed by using TA Instruments equipment model DSC Q100 in the range from $-40\ ^\circ$C to $45\ ^\circ$C at a heating rate of $10\ ^\circ$C/min.

2.2.6. Mechanical Tests

The tensile mechanical properties of the developed elastomer composites were analyzed by means of an MTS Criterion Model 43 equipped with a 1 kN load cell. The used samples had a dog bone shape (type 1) with a thickness between 2 and 3 mm, a length of 115 mm and a width in the narrower section of 6.2 ± 0.2 mm. The tensile tests were carried out at a crosshead speed of 50 mm/min until the sample broke, according to standard ISO 37:2013. The resilience tests were carried out by performing four tensile test cycles from zero load to a deformation of 14% and returning to zero load.

The Shore A hardness of the elastomers was measured with a Sauter HDA100-1 tester.

All the thermal and mechanical characterization tests were performed three times.

3. Results and Discussion

3.1. Elastomer Synthesis

A total of 10 different elastomers were produced as result of varying the [NCO]/[OH] molar ratio and the mSD-(LDPE·EVA-RT27) content.

The composites were synthesized with [NCO]/[OH] molar ratios of 1.05 and 1.1 while varying the mSD-(LDPE·EVA-RT27) content within 0.0–20.0 wt.%. The composite names include all this information by putting after the letter E (corresponding to elastomer) the [NCO]/[OH] molar ratio and the mSD-(LDPE·EVA-RT27) content, separated by dashes. The names and values of both variables for the different synthesized elastomers are gathered in Table 1.

Table 1. [NCO]/[OH] molar ratios and mSD-(LDPE·EVA-RT27) contents of the synthesized elastomers.

Name	[NCO]/[OH]	mSD-(LDPE·EVA-RT27) (wt.%)
E-1.05-0		0.0
E-1.05-5		5.0
E-1.05-10	1.05	10.0
E-1.05-15		15.0
E-1.05-20		20.0
E-1.1-0		0.0
E-1.1-5		5.0
E-1.1-10	1.1	10.0
E-1.1-15		15.0
E-1.1-20		20.0

All the formulations allowed successful synthesis of the elastomers, as can be seen in Figure 1.

At a simple view, the elastomers reagents successfully reacted. However, in order to confirm the formation of the characteristics of the elastomers' structures, FTIR analyses of the composites with different mSD-(LDPE·EVA-RT27) contents and [NCO]/[OH] molar ratios were carried out (Figure 2).

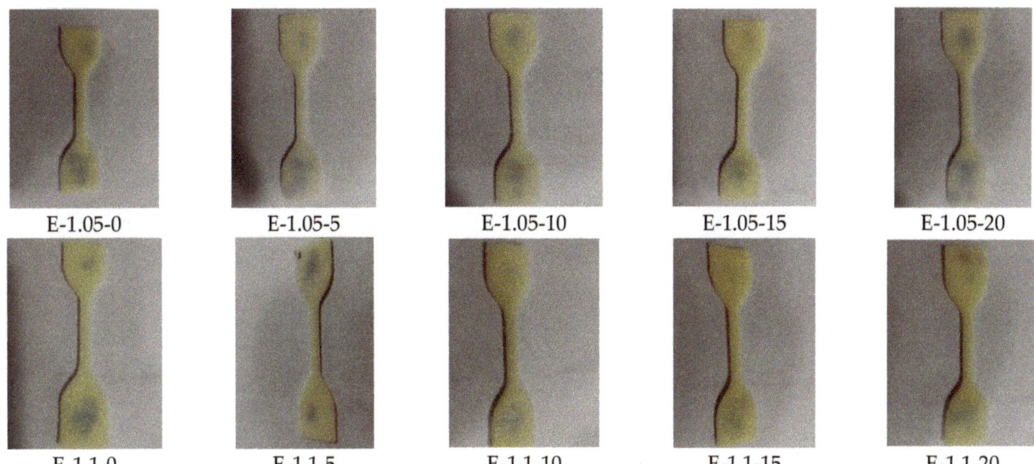

E-1.05-0   E-1.05-5   E-1.05-10   E-1.05-15   E-1.05-20

E-1.1-0   E-1.1-5   E-1.1-10   E-1.1-15   E-1.1-20

Figure 1. Appearance of the synthesized elastomers with different [NCO]/[OH] and mSD-(LDPE·EVA-RT27) contents.

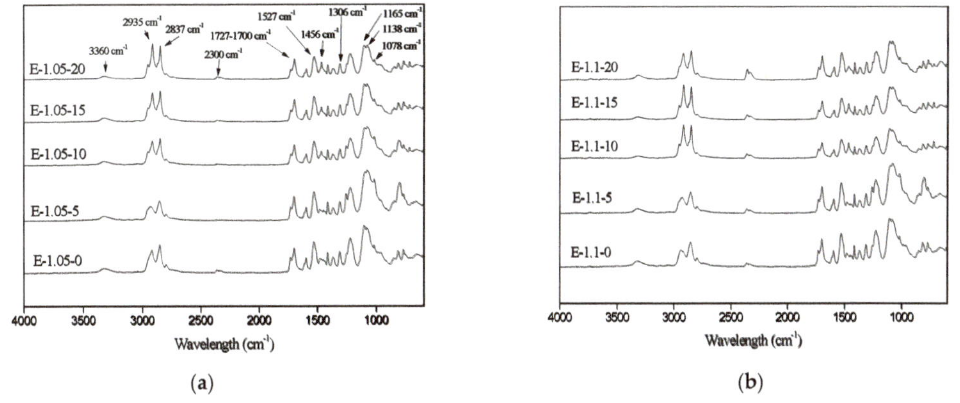

(a)   (b)

Figure 2. FT-IR spectrums of the synthesized elastomers with different mSD-(LDPE·EVA-RT27) contents and [NCO]/[OH] values of (**a**) 1.05 and (**b**) 1.1.

The presence of urethane groups in the synthesized elastomers can be clearly identified in Figure 2 by the stretching vibrations of the -NH bonds around 3360 cm^{-1} [17]. Aside from that, the characteristic signals of the carbonyl groups (-C=O), hydrogen bonds, -NH bending vibrations and -CN stretching vibrations appeared at 1727, 1700, 1527 and 1306 cm^{-1}, respectively [18]. The -NH groups could join the -C=O by the hydrogen bonds in the hard segments and with the -C-O-C of the soft segments. The peaks at 2935 cm^{-1} and 2837 cm^{-1} correspond to the symmetric and asymmetric stretching vibrations of -CH$_2$ group, respectively, while at 1373 cm^{-1} and 1456 cm^{-1}, bending stretching in the polyol chain appeared [19]. In addition, the lack of significant stretching vibrations of the isocyanate group (N=C=O) of MDI at 2300 cm^{-1} confirmed the success of the reaction for the case of a [NCO]/[OH] molar ratio of 1.05 [20]. On the contrary, that signal appeared for the case of a [NCO]/[OH] molar ratio of 1.1, indicating the excess of isocyanate groups for the case of this formulation. On the other hand, the existence of a peak at 1594 cm^{-1} indicated the presence of aromatic groups coming from MDI [18]. Finally, the peaks that appeared at 1165, 1138 and at 1078 cm^{-1} corresponded to the stretching and bending vibrations of C-O-C present in the ether, urethane and ester groups of the elastomers, respectively [18].

3.2. Elastomer Structure and Filler Distribution

Once the proper chemical structures of the synthesized elastomers were confirmed, their morphologies were observed by SEM for further understanding of their physical and mechanical properties. Figure 3 shows two examples of the typical filler distribution in matrix that were very similar in the rest of the samples, and Figure 4 shows the SEM pictures of the different synthesized elastomer matrices.

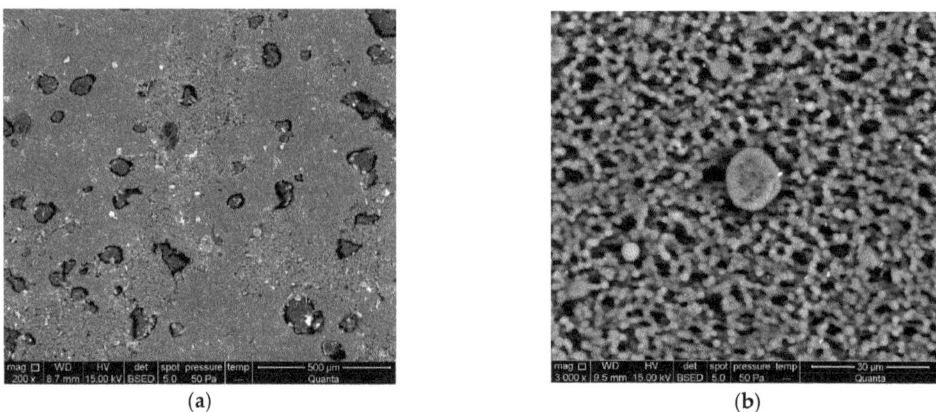

(**a**) (**b**)

Figure 3. SEM photos of filler distributions for the E-1.1-20 at 200× magnification (**a**) and for the E-1.05-15 at 3000× magnification (**b**).

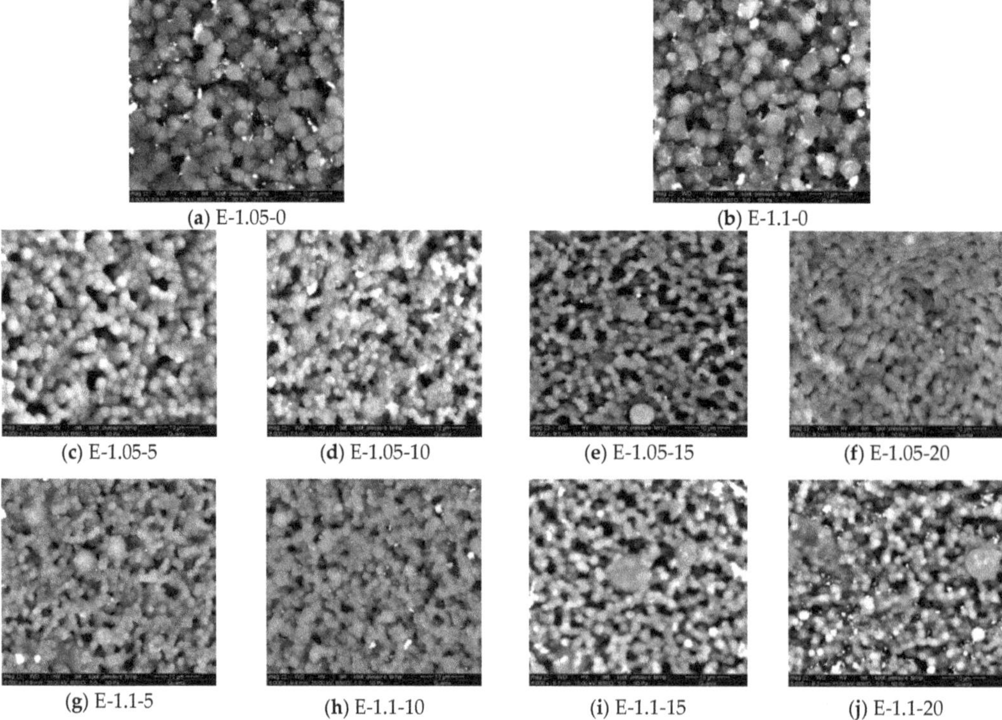

(**a**) E-1.05-0 (**b**) E-1.1-0

(**c**) E-1.05-5 (**d**) E-1.05-10 (**e**) E-1.05-15 (**f**) E-1.05-20

(**g**) E-1.1-5 (**h**) E-1.1-10 (**i**) E-1.1-15 (**j**) E-1.1-20

Figure 4. SEM pictures at 6000× magnification of the synthesized elastomer matrices with different mSD-(LDPE·EVA-RT27) contents and [NCO]/[OH] values of 1.05 and 1.1.

The filler distribution in the polymer matrix was not totally homogeneous, as can be seen in Figure 3. Many of the microcapsules were well distributed, as shown in Figure 3b, and some others were agglomerated, as can be seen in Figure 3a. The rest of the synthesized elastomers also presented mSD-(LDPE·EVA-RT27) distributed for the whole polymer matrix with some agglomeration points.

As can be seen in Figure 4, the elastomers presented a structure of polyurethanes derived from aromatic diisocyanates consisting of a net of spherical microparticles [21,22]. Aside from that, most particles presented a homogeneous size, except for some bigger particles. The most accurate study of the influence of the mSD-(LDPE·EVA-RT27) content and of the [NCO]/[OH] molar ratio on the particle size was carried out by using Motic Plus software (see Table 2).

Table 2. Average particle size of the elastomer polymeric matrices.

Elastomer	Average Particle Size (μm)	Maximum Value(μm)	Minimum Value(μm)
E-1.05-0	2.640 $\pm$ 0.409	3.689	0.785
E-1.05-5	2.049 $\pm$ 0.227	3.759	0.689
E-1.05-10	1.735 $\pm$ 0.210	2.388	1.271
E-1.05-15	1.617 $\pm$ 0.283	2.288	0.684
E-1.05-20	1.896 $\pm$ 0.346	2.537	1.022
E-1.1-0	2.220 $\pm$ 0.696	4.862	1.449
E-1.1-5	1.893 $\pm$ 0.212	2.480	1.473
E-1.1-10	1.693 $\pm$ 0.193	2.230	1.201
E-1.1-15	1.868 $\pm$ 0.234	2.483	1.164
E-1.1-20	1.734 $\pm$ 0.295	3.110	1.081

According to the results, the addition of mSD-(LDPE·EVA-RT27) reduced the particle size of the polymeric matrix and improved the sample structure homogeneity, since lower standard deviations were obtained. The particle size of the polymeric matrix structure was at a minimum for E-1.05-15 and E-1.1-10. Thus, the presence of MC increased the nucleation point during polymer formation, reducing the particle size of the polymeric matrix up to 10–15 wt.%. Nevertheless, from this point, the effect of the agglomeration due to the increase of the sample viscosity stabilized or even increased the particle size of the polymeric matrix [23].

On the other hand, the effect of the elastomer structure on the porosity and density was also analyzed, and both effects are shown in Figure 5.

As can be seen, the higher the MC content, the lower the sample porosity, probably due to the reduction of the particle size of the polymeric matrix, the presence of the MC in the structure and the increase of the reactive mixture viscosity. On the other hand, despite the porosity decrease, the density also decreased, since the mSD-(LDPE·EVA-RT27) density was 0.866 g/cm^3 [24], 26.7% lower than the net elastomer density. In the case of elastomer E-1.05, the density decrease was almost linearly dependent on the mSD-(LDPE·EVA-RT27) content. A similar tendency was observed up to an MC content of 10 wt.% for the elastomer E-1.1. Nevertheless, from that point, higher densities than those obtained for the composites from a [NCO]/[OH] molar ratio of 1.05 were observed, possibly due to the excess of isocyanate.

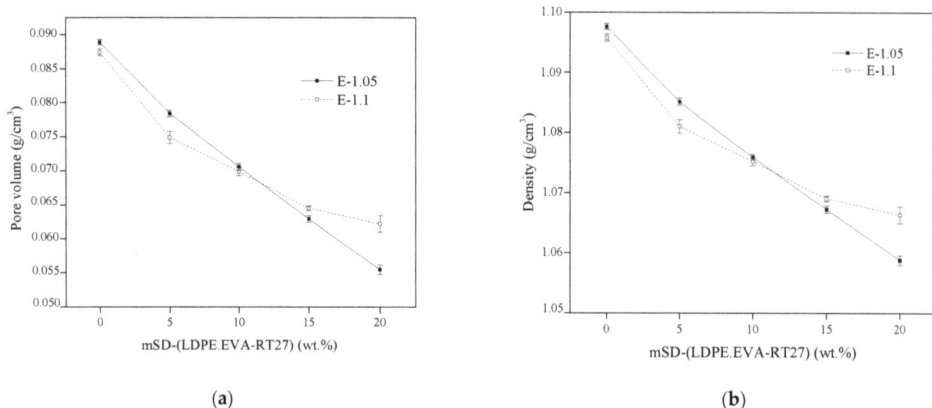

(a) (b)

Figure 5. Porosity (**a**) and density (**b**) of the elastomers with different mSD-(LDPE·EVA-RT27) contents and [NCO]/[OH] molar ratios.

3.3. Thermal Stability

The thermal stability of PU elastomers depends on multiple factors, such as the proportion between soft and hard segments, the chain extensor type and the [NCO]/[OH] molar ratio [18,20,25]. Figure 6 shows the thermogravimetric results of the mSD-(LDPE·EVA-RT27) and pure paraffin (Rubitherm®RT27) used in the microcapsules' synthesis and all the synthesized elastomers.

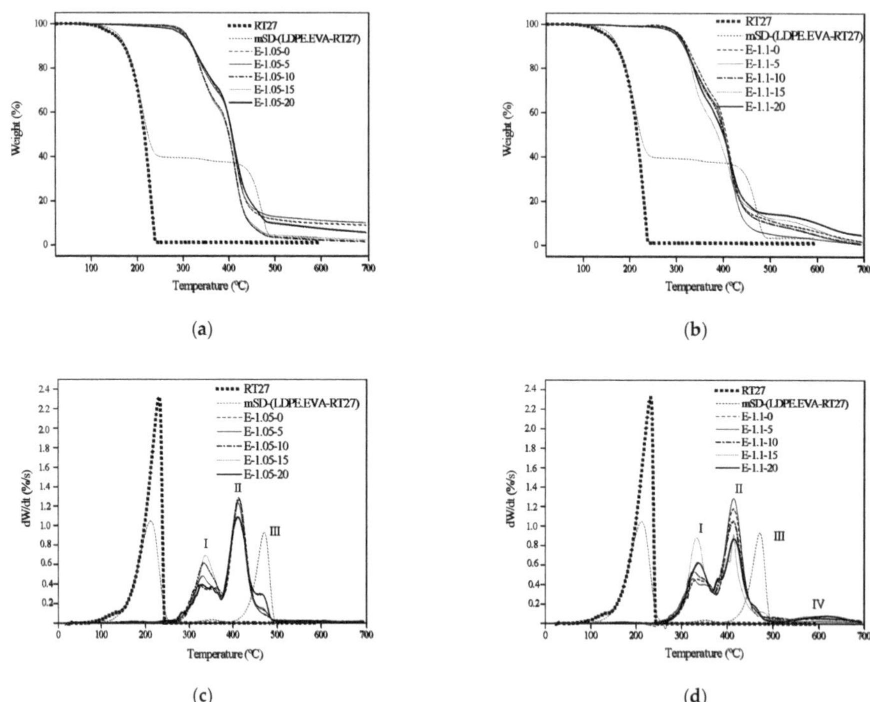

(a) (b)

(c) (d)

Figure 6. Thermogravimetric analyses (**a**,**b**) and their DTGs (**c**,**d**) of the synthesized elastomers, mSD-(LDPE·EVA-RT27) and the pure paraffin. Peaks I = HS degradation; Peak II = SS degradation; Peak III = high molecular weight reticulated segment degradation; Peak VI = allophanates degradation.

As can be seen in Figure 6, the pure paraffin and the paraffin from mSD-(LDPE·EVA-RT27) were completely evaporated at 210 °C, whereas the shell material of the microcapsules was degraded at 480 °C. On the other hand, the thermal degradation of the polyurethane elastomers was a complex process with three different thermal degradation steps. The first degradation step corresponding with urethane bond decomposition from the hard segments (peak I of the DTG elastomer curves) occurred between 300 and 380 °C. The second degradation step appeared between 380 and 500 °C, and this can be attributed to the soft segment degradation (peak II). Finally, there was a third small step from 550 °C, which was due to the presence of reticulated segments of a high molecular weight. It is worth pointing out that all the elastomers were stable up to 280 °C, and therefore, the paraffin evaporation was delayed, probably evaporating while the urethane bond degradation took place. This could be due to the immersion of the microcapsules in the polymer matrix acting as a protected barrier, forming a form-stable material. Aside from that, it can be noticed that for the [NCO]/[OH] molar ratio of 1.05, there was a residue with a maximum value of 10 wt.% for the case without PCMs. On the contrary, elastomers from the [NCO]/[OH] molar ratio of 1.1 presented a fourth degradation peak and absence of residue, probably because of the excess of -NCO reacting with the urethane groups at high temperatures and forming allophanates that decomposed around 600 °C [26].

From the TGA results, the experimental weight loss corresponding to the soft and hard segments of the elastomers (SS and HS, respectively) and the mSD-(LDPE·EVA-RT27) shell for each one of the synthesized elastomers was determined and compared to the theoretical expected values. Figure 7 shows the experimental and theoretical weight loss for the SS, HS and mSD-(LDPE·EVA-RT27) shells, depending on the microcapsules content and the [NCO]/[OH] molar ratio. The theoretical values were estimated from the mass contents of the microcapsules, their compositions and the mass of polyurethane components added in the elastomer synthesis recipe.

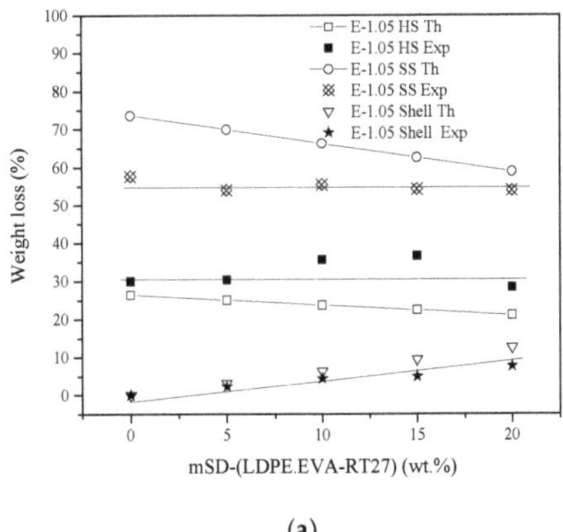

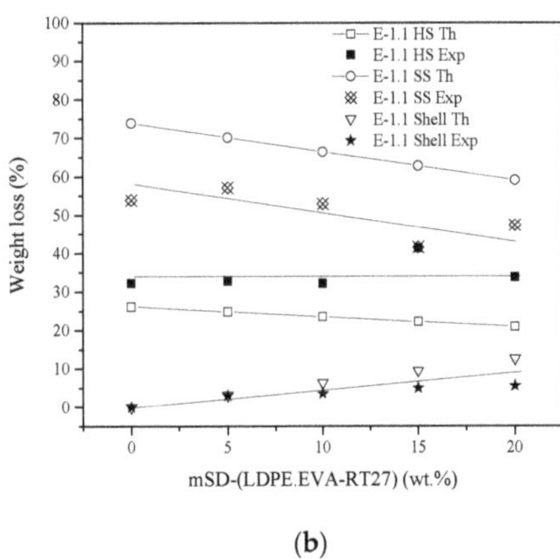

Figure 7. Experimental and theoretical weight loss for the SS, HS and microcapsule shells, depending on the mSD-(LDPE·EVA-RT27) content, for [NCO]/[OH] molar ratios of (**a**) 1.05 and (**b**) 1.1. The straight lines indicate the linear trends of the weight loss.

It can be observed in Figure 7 that the HS proportion in the elastomers increased with the microcapsule content up to the case of 15 wt.% mSD-(LDPE·EVA-RT27), probably due to the fact that paraffin degradation also occurred at this degradation step, as was mentioned before. However, for the case of 20 wt.% mSD-(LDPE·EVA-RT27), the elastomers presented a HS percentage value similar to that of the pure elastomers. This can be attributed to poor homogenous elastomers, in which the particles would tend to form agglomerates. Aside from that, the HS experimental contents were higher than the theoretical ones. Again, this was partially due to the paraffin degradation in that step. On the contrary, the SS weight loss values were lower than expected. Therefore, the variation in both segments' contents could be due to a partial overlap between the degradation steps that hampered the quantification of the segment contents considering part of the SS in the HS. On the other hand, in the case of the SS weight loss, it remained nearly constant independent of the MC content, and it was lower for the case of a [NCO]/[OH] value of 1.1. The lower values with the increase of the NCO content could be due to the formation of higher molecular weight structures that degraded at higher temperatures [27,28]. This is in agreement with the appearance of a fourth degradation step from 550 °C for the case of a [NCO]/[OH] value of 1.1.

As expected, the higher the MC content, the higher the weight loss corresponding to the shell. The deviation between the experimental and theoretical values of the shell weight loss also increased with the microcapsule content, being especially significant (46%) for the case of a 20 wt.% content. Again, the elastomer with a 20 wt.% content presented different behavior, probably due to the MC agglomeration and subsequent heterogeneity of the polymer matrix.

3.4. Latent Heat

The final aim of the mSD-(LDPE·EVA-RT27) incorporation was to increase the thermal energy storage capacity of the PU elastomers. Thus, the latent heat at different points of the different synthesized elastomers was determined via DSC. The average latent heat (ΔH_m), the theoretical latent heat (ΔH_t), the standard deviation and the thermal yield are gathered in Table 3.

Table 3. Average latent heat (ΔH_m), theoretical latent heat (ΔH_t), standard deviation, melting temperature (T_m), crystallization temperature (T_c) and yield of the elastomers with different mSD-(LDPE·EVA-RT27) contents and [NCO]/[OH] molar ratios.

	[NCO]/[OH] 1.05										[NCO]/[OH] 1.1	
MC (%)	ΔH_m (J/g)	ΔH_t (J/g)	T_m (°C)	T_c (°C)	Standard Deviation	Yield (%)	ΔH_m (J/g)	ΔH_t (J/g)	T_m (°C)	T_c (°C)	Standard Deviation	Yield (%)
mSD-(LDPE·EVA-RT27)			22.92	17.38					22.92	17.38		
5	1.92	2.72	23.29	22.78	1.26	70.44	1.48	2.72	24.10	23.79	1.09	54.40
10	4.25	5.44	25.61	24.06	2.43	78.02	4.18	5.44	25.04	21.55	2.47	76.82
15	7.39	8.17	26.15	22.36	1.62	90.51	6.53	8.17	26.70	22.36	2.61	79.95
20	7.52	10.89	24.84	18.38	2.02	69.09	8.70	10.89	26.09	19.81	2.28	79.90

The net elastomers did not show any peak related to a melting or solidifying point, contrary to that observed for those containing mSD-(LDPE·EVA-RT27). As expected, the higher the mSD-(LDPE·EVA-RT27) content, the higher the latent heat of the elastomers. It can be also observed that the standard deviations were lower for the case of elastomer E-1.05, indicating a better distribution of the microcapsules into the elastomers and exhibiting a large yield, except for the 20 wt.% microcapsules. Taking into account the yield and the standard deviation, the elastomeric matrix from a [NCO]/[OH] molar ratio of 1.05 was not able to stabilize the total amount of MC, attending to the lower porosity exhibited when the 20 wt.% microcapsules case was used. Hence, although composites from the [NCO]/[OH] molar ratios of 1.1 up to 15 wt.% microcapsules presented lower latent heats than those corresponding materials with a [NCO]/[OH] molar ratio of 1.05, the elastomer with the highest thermoregulating capacity was E-1.1-20, exhibiting a latent heat of 8.7 J/g. This value was 15.69% higher than the corresponding E-1.05-20 elastomer. Hence, although it was observed in the structural and stability tests that the higher content of NCO groups, the higher the reticulation and hampering of good microcapsule distribution up to 15 wt.% microcapsules, the porosity and the thermoregulating properties of the

E-1.1-20 elastomer were higher than the corresponding properties of the E-1.05-20. These better properties could be attributed to the lower average particle size of this elastomer matrix (1.734 mm) with respect to that of E-1.05-20 (1.896 mm), which also led to a large density ($r_{E\text{-}1.1\text{-}20} > r_{E\text{-}1.05\text{-}20}$; $1.07 > 1.06$ g/cm^3). Finally, there was a slight increase in the melting points for all composites as result of the porosity and density diminution of the microcapsule contents, causing a reduction of the global thermal conductivity of these form-stable materials. On the other hand, the crystallization temperature was practically equal to the melting point for microcapsule contents lower than 15 wt.%, indicating that these form-stable materials did not exhibit any hysteresis. Nevertheless, for microcapsule contents within 15–20 wt.%, the crystallization temperature tended toward that of the original microcapsules, indicating that they worked as the key thermal material instead of the matrix elastomer.

3.5. Mechanical Properties

As was commented on in the characterization techniques section, the tensile stress–strain properties were determined following the standard UNE-ISO 37:2013. The tensile strength at break and at a deformation of 150% are important properties for elastomers. The values obtained for the different synthesized materials are shown in Figure 8.

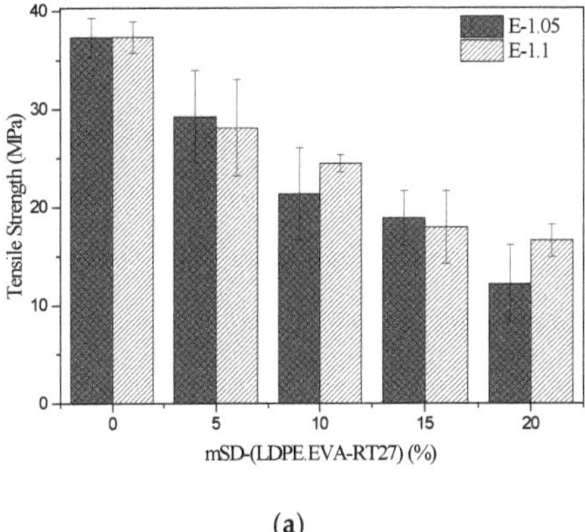

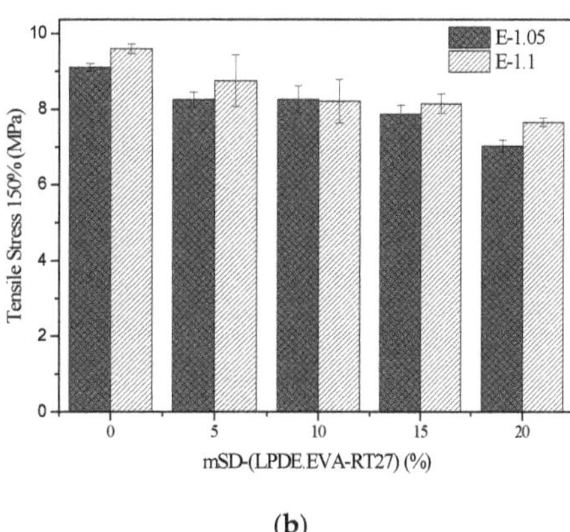

Figure 8. The tensile strength at break (**a**) and tensile stress at a deformation of 150% (**b**).

A reduction of both properties when the mSD-(LDPE·EVA-RT27) microcapsules were added can be observed, especially for E-1.05-20. This decrease could be caused by the elastomer heterogeneity promoted by worse microcapsule distribution, which favors the fractured appearance. On the other hand, elastomer E-1.1 presented better results than E-1.05 in the case of tensile stress at a deformation of 150%, suggesting greater rigidity of E-1.1, while the tensile strength results were very similar.

Another important mechanical property is the resilience, since it gives an idea of the deformation energy per volume unit that can be recovered after the material's deformation. This refers to the polymer's capacity for recovering after a deformation. The resilience of the synthesized elastomers is shown in Figure 9a.

According to the experimental results, the presence of the microcapsules improved the resilience of the elastomers, especially for the case of E-1.05. Nevertheless, the resilience of elastomer E-1.1 was always better than that observed for elastomer E-1.05. It is worth mentioning that the maximum resilience was found independently of the [NCO]/[OH] molar ratio for an mSD-(LDPE·EVA-RT27) content of 5 wt.%.

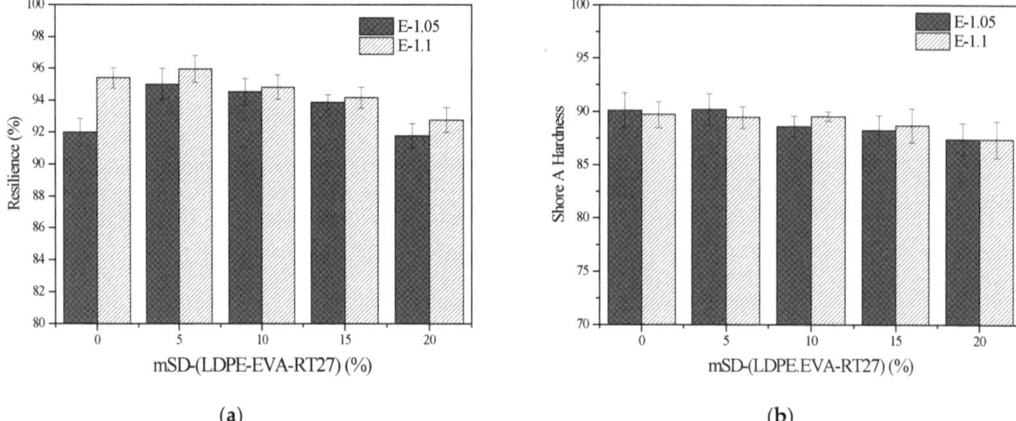

Figure 9. Resilience (**a**) and Shore A hardness (**b**) of the elastomers with mSD-(LDPE·EVA-RT27).

The hardness of the elastomers was tested with a Shore A durometer, and the values are shown in Figure 9b. It can be observed that the hardness was not influenced by the [NCO]/[OH] molar ratio, as the results were very similar for composites E-1.05 and E-1.1. Aside from that, the addition of microcapsules did not change the hardness of the elastomers significantly, with a maximum decrease of 2.98% observed for elastomers containing 20 wt.% MC.

4. Conclusions

Elastomeric materials with thermoregulating properties due to incorporating mSD-(LDPE·EVA-RT27) within 0–20 wt.% and isocyanates/hydroxyl [NCO]/[OH] molar ratios of 1.05 and 1.1 were produced.

The presence of urethane and isocyanate groups was clearly identified by FT-IR analyses, allowing the confirmation of an excess of isocyanate groups for the formulated [NCO]/[OH] molar ratio 1.1.

All the synthesized elastomers were formed by a net of spherical microparticles with an average particle size within 2.640–1.617 mm.

Thermogravimetric analyses allowed for distinguish between weight losses attributed to paraffin, the microcapsule shells and the hard segments and soft segments. Nevertheless, the weight loss of paraffin in the elastomers was overlayed with the hard segment, indicating that the elastomers were able to build form-stable materials, enlarging the paraffin thermal stability since its vaporization was delayed.

In general, the density and mechanical properties decreased with the microcapsule content as result of the lower density of the microcapsules and their worse distribution into the elastomeric matrix. Elastomer E-1.05 exhibited the better structural and stability properties for MC contents up to 15 wt.%, whereas the E-1.1 elastomer containing 20 wt.% MC revealed mechanical and thermal synergy effects, demonstrating good structural stability and the largest latent heat (8.7 J/g).

Author Contributions: Conceptualization, I.I., Á.S., M.C. and J.F.R.; methodology, P.J.T., A.M.B., J.D., M.C. and I.G.; validation, Á.S., P.J.T., J.D. and I.G.; formal analysis, A.M.B., I.I., Á.S., M.C. and J.F.R.; investigation, Á.S., I.I. and P.J.T.; resources, M.C. and J.F.R.; data curation, A.M.B. and I.I.; writing—original draft preparation, I.I. and A.M.B.; writing—review and editing, I.G. and M.C.; visualization, I.I. and A.M.B.; supervision, J.F.R. and M.C.; project administration, M.C.; funding acquisition, M.C., and J.F.R. All authors have read and agreed to the published version of the manuscript.

Funding: This research was funded by the Spanish Ministry of Science, Innovation and Universities under project TRANSENERGY (RTI2018-100745-B-100).

Institutional Review Board Statement: Not applicable.

Informed Consent Statement: Not applicable.

Data Availability Statement: Data can be supplied by e-mail if they are required.

Conflicts of Interest: The authors declare no conflict of interest.

References

1. Polyurethane Market Size Worth $105.2 Billion by 2025 | CAGR: 7.0%. Available online: https://www.grandviewresearch.com/press-release/global-polyurethane-pu-market (accessed on 5 April 2021).
2. Simón, D.; Borreguero, A.; de Lucas, A.; Rodríguez, J. Recycling of polyurethanes from laboratory to industry, a journey towards the sustainability. *Waste Manag.* **2018**, *76*, 147–171. [CrossRef] [PubMed]
3. Simón, D.; Borreguero, A.M.; de Lucas, A.; Gutiérrez, C.; Rodríguez, J.F. *Sustainable Polyurethanes: Chemical Recycling to Get It*; Springer: Cham, Switzerland, 2014; pp. 229–260.
4. Xie, F.; Zhang, T.; Bryant, P.; Kurusingal, V.; Colwell, J.M.; Laycock, B. Degradation and stabilization of polyurethane elastomers. *Prog. Polym. Sci.* **2019**, *90*, 211–268. [CrossRef]
5. Crawford, D.M.; Bass, R.G.; Haas, T.W. Strain effects on thermal transitions and mechanical properties of thermoplastic polyurethane elastomers. *Thermochim. Acta* **1998**, *323*, 53–63. [CrossRef]
6. Mondal, S.; Das, A.; Bandyopadhyay, J.; Ray, S.S.; Heinrich, G.; Bandyopadhyay, A. A noble additive cum compatibilizer for dispersion of nanoclay into ethylene octene elastomer. *Appl. Clay Sci.* **2016**, *126*, 41–49. [CrossRef]
7. Zare, Y. Estimation of material and interfacial/interphase properties in clay/polymer nanocomposites by yield strength data. *Appl. Clay Sci.* **2015**, *115*, 61–66. [CrossRef]
8. Annamalai, P.K.; Dagnon, K.L.; Monemian, S.; Foster, E.J.; Rowan, S.J.; Weder, C. Water-Responsive Mechanically Adaptive Nanocomposites Based on Styrene–Butadiene Rubber and Cellulose Nanocrystals—Processing Matters. *ACS Appl. Mater. Interfaces* **2013**, *6*, 967–976. [CrossRef]
9. Han, D.; Qu, M.; Yue, C.Y.; Lou, Y.; Musso, S.; Robisson, A. Swellable elastomeric HNBR–MgO composite: Magnesium oxide as a novel swelling and reinforcement filler. *Compos. Sci. Technol.* **2014**, *99*, 52–58. [CrossRef]
10. Sarier, N.; Onder, E. Thermal insulation capability of PEG-containing polyurethane foams. *Thermochim. Acta* **2008**, *475*, 15–21. [CrossRef]
11. Cannon, F.R.; Denhartog, E.A. Quantitative evaluation of mattresses using a thermal seat tester. *J. Text. Inst.* **2019**, *110*, 1352–1358. [CrossRef]
12. Giro-Paloma, J.; Martínez, M.; Cabeza, L.F.; Fernández, A.I. Types, methods, techniques, and applications for microencapsulated phase change materials (MPCM): A review. *Renew. Sustain. Energy Rev.* **2016**, *53*, 1059–1075. [CrossRef]
13. Borreguero, A.M.; Rodríguez, J.F.; Valverde, J.L.; Peijs, T.; Carmona, M. Characterization of rigid polyurethane foams containing microencapsulted phase change materials: Microcapsules type effect. *J. Appl. Polym. Sci.* **2013**, *128*, 582–590. [CrossRef]
14. Armstrong, D.P.; Chatterjee, K.; Ghosh, T.K.; Spontak, R.J. Form-stable phase-change elastomer gels derived from thermoplastic elastomer copolyesters swollen with fatty acids. *Thermochim. Acta* **2020**, *686*, 178566. [CrossRef]
15. Juárez, D.; Ferrand, S.; Fenollar, O.; Fombuena, V.; Balart, R. Improvement of thermal inertia of styrene–ethylene/butylene–styrene (SEBS) polymers by addition of microencapsulated phase change materials (PCMs). *Eur. Polym. J.* **2011**, *47*, 153–161. [CrossRef]
16. Chriaa, I.; Trigui, A.; Karkri, M.; Jedidi, I.; Abdelmouleh, M.; Boudaya, C. Thermal properties of shape-stabilized phase change materials based on Low Density Polyethylene, Hexadecane and SEBS for thermal energy storage. *Appl. Therm. Eng.* **2020**, *171*, 115072. [CrossRef]
17. Simón, D.; De Lucas, A.; Rodríguez, J.; Borreguero, A. Glycolysis of high resilience flexible polyurethane foams containing polyurethane dispersion polyol. *Polym. Degrad. Stab.* **2016**, *133*, 119–130. [CrossRef]
18. Włoch, M.; Datta, J. Synthesis, Structure and Properties of Poly(ester-Urethane-Urea)s Synthesized Using Biobased Diamine. *J. Renew. Mater.* **2016**, *4*, 72–77. [CrossRef]
19. Simón, D.; Borreguero, A.; de Lucas, A.; Rodríguez, J. Valorization of crude glycerol as a novel transesterification agent in the glycolysis of polyurethane foam waste. *Polym. Degrad. Stab.* **2015**, *121*, 126–136. [CrossRef]
20. Rao, Y.; Munro, J.; Ge, S.; Garcia-Meitin, E. PU elastomers comprising spherical nanosilicas: Balancing rheology and properties. *Polymer* **2014**, *55*, 6076–6084. [CrossRef]
21. Taheri, N.; Sayyahi, S. Effect of clay loading on the structural and mechanical properties of organoclay/HDI-based thermoplastic polyurethane nanocomposites. *e-Polymers* **2016**, *16*, 65–73. [CrossRef]
22. Prisacariu, C. *Polyurethane Elastomers From Morphology to Mechanical Aspects*; Springer: Berlin/Heidelberg, Germany, 2011; p. 414.
23. Ziegmann, A.; Schubert, D.W. Influence of the particle size and the filling degree of barium titanate filled silicone elastomers used as potential dielectric elastomers on the mechanical properties and the crosslinking density. *Mater. Today Commun.* **2018**, *14*, 90–98. [CrossRef]
24. Serrano, A.; Borreguero, A.M.; Garrido, I.; Rodríguez, J.F.; Carmona, M. The role of microstructure on the mechanical properties of polyurethane foams containing thermoregulating microcapsules. *Polym. Test.* **2017**, *60*, 274–282. [CrossRef]

25. Mahmood, K.; Zia, K.M.; Aftab, W.; Zuber, M.; Tabasum, S.; Noreen, A.; Zia, F. Synthesis and characterization of chitin/curcumin blended polyurethane elastomers. *Int. J. Biol. Macromol.* **2018**, *113*, 150–158. [CrossRef] [PubMed]
26. Lapprand, A.; Boisson, F.; Delolme, F.; Mechin, F.; Pascault, J.-P. Reactivity of isocyanates with urethanes: Conditions for allophanate formation. *Polym. Degrad. Stab.* **2005**, *90*, 363–373. [CrossRef]
27. Poljanšek, I.; Fabjan, E.; Moderc, D.; Kukanja, D. The effect of free isocyanate content on properties of one component urethane adhesive. *Int. J. Adhes. Adhes.* **2014**, *51*, 87–94. [CrossRef]
28. Daniel-Da-Silva, A.L.; Bordado, J.C.M.; Martín-Martínez, J.M. Evidences of phase separation in moisture-cured poly(urethane urea)s by means of temperature modulated differential scanning calorimetry. *J. Polym. Sci. Part B Polym. Phys.* **2007**, *45*, 3034–3045. [CrossRef]

applied sciences

Article

Storage Capacity in Dependency of Supercooling and Cycle Stability of Different PCM Emulsions

Stefan Gschwander *, Sophia Niedermaier, Sebastian Gamisch, Moritz Kick , Franziska Klünder and Thomas Haussmann

Fraunhofer Institute for Solar Energysystems, Heidenhofstr. 2, 79110 Freiburg, Germany; sophia.niedermaier@ise.fraunhofer.de (S.N.); sebastian.gamisch@ise.fraunhofer.de (S.G.); moritz.kick@ise.fraunhofer.de (M.K.); franziska.kluender@ise.fraunhofer.de (F.K.); thomas.haussmann@ise.fraunhofer.de (T.H.)
* Correspondence: Stefan.gschwander@ise.fraunhofer.de

Abstract: Phase-change materials (PCM) play off their advantages over conventional heat storage media when used within narrow temperature ranges. Many cooling and temperature buffering applications, such as cold storage and battery cooling, are operated within small temperature differences, and therefore, they are well-suited for the application of these promising materials. In this study, the storage capacities of different phase-change material emulsions are analysed under consideration of the phase transition behaviour and supercooling effect, which are caused by the submicron size scale of the PCM particles in the emulsion. For comparison reasons, the same formulation for the emulsions was used to emulsify 35 wt.% of different paraffins with different purities and melting temperatures between 16 and 40 °C. Enthalpy curves based on differential scanning calorimeter (DSC) measurements are used to calculate the storage capacities within the characteristic and defined temperatures. The enthalpy differences for the emulsions, including the first phase transition, are in a range between 69 and 96 kJ/kg within temperature differences between 6.5 and 10 K. This led to an increase of the storage capacity by a factor of 2–2.7 in comparison to water operated within the same temperature intervals. The study also shows that purer paraffins, which have a much higher enthalpy than blends, reveal, in some cases, a lower increase of the storage capacity in the comparison due to unfavourable crystallisation behaviour when emulsified. In a second analysis, the stability of emulsions was investigated by applying 100 thermal cycles with defined mechanical stress at the same time. An analysis of the viscosity, particle size and melting crystallisation behaviour was done by showing the changes in each property due to the cycling.

Keywords: phase-change material; dispersion; thermal-mechanical stability; viscosity; supercooling; nucleating agent; cold storage; battery cooling

Citation: Gschwander, S.; Niedermaier, S.; Gamisch, S.; Kick, M.; Klünder, F.; Haussmann, T. Storage Capacity in Dependency of Supercooling and Cycle Stability of Different PCM Emulsions. *Appl. Sci.* **2021**, *11*, 3612. https://doi.org/10.3390/app11083612

Academic Editor: Ioannis Kartsonakis

Received: 18 March 2021
Accepted: 12 April 2021
Published: 16 April 2021

Publisher's Note: MDPI stays neutral with regard to jurisdictional claims in published maps and institutional affiliations.

1. Introduction

PCM emulsions or, more generally, PCM slurries (PCS) are heat transfer fluids consisting of a PCM that is dispersed in a fluid (in this study, water). Compared to pure water, they showed a high storage capacity due to the usage of the PCM melting and solidification process, in addition to the sensible heat capacity of the water and the PCM. To utilise this benefit, they have to be used within the PCM melting and crystallisation temperature range [1]. This offers the advantage of reduced volumes for energy storage due to a greater storage density [2]. PCS, in general, offer the advantage of being liquid-independent of the PCM's state of matter. This property makes it possible to convey PCS in hydraulic systems equally in both physical states of the PCM. Hence, PCS can act as heat transfer fluids and as a storage fluid. With this functionality, PCS contributes as a storage material to thermal systems that are operated within small temperature differences, such as cooling or temperature-buffering applications. PCS can transfer this heat homogeneously at an almost constant temperature [3]. The fine dispersion of the PCM in water leads to a high surface-to-volume ratio that contributes to a high heat transfer rate between water and PCM [4,5].

Over time, researchers have investigated different paraffins for use as PCM in emulsions. Already, in the 1990s, Inaba and Morita [6] investigated the application of tetradecane with a melting temperature of 14.75 °C in water as a PCM emulsion for cold storage. A mixture of anionic and non-ionic emulsifiers stabilised the emulsion. Paraffins are available over a wide range of melting temperatures. Additionally, melting temperatures between that of pure paraffins can be achieved by mixing two different paraffins [7]. He and Setterwall [7] mixed, for example, n-tetradecane and n-hexadecane and found melting temperatures between those of the pure materials. They also found that a mixture of about 90 mol% n-tetradecane and 10 mol% n-hexadecane showed a melting temperature at 2 °C. This makes it possible to adapt the melting point to the demand of different applications by choosing a paraffin with an appropriate melting temperature or mixed paraffins with higher and lower melting temperatures.

According to Montenegro and Landfester [8], n-alkanes with an even number of carbon atoms in the range from 12 to 26 crystallise in the triclinic form, and those with an odd-numbered carbon chain show a phase transition from liquid to a metastable orthorhombic rotator phase and, in the second step, at a lower temperature, a transition to the stable phase. It was also observed that even-numbered n-alkanes show a phase transition to the rotator phase when emulsified in the nanometre scale. Hagelstein und Gschwander [9] observed this as well for octadecane emulsions. It was found that the transition to a rotator phase accounts for 70–75% of the total crystallisation enthalpy. Beside this change in the phase transition, it was observed that, due to the confined volume of the paraffin droplets, it was crystallising at a temperature below its bulk phase-change temperature from solid to liquid. This is due to a separation of the natural given crystallisation seeds [10]. This hysteresis effect is called supercooling. Zhang et al. [11] reported that a PCM emulsion with a melting temperature of 18 °C had to be cooled down to 7 °C to obtain the full latent heat. Nucleation agents were used to supress the supercooling. Huang et al. [5] developed, for example, a PCM emulsions based on the paraffin blends RT6 with a melting temperature of 6 °C, RT10 (10 °C) and RT20 (20 °C) from Rubitherm Technologies GmbH. Non-ionic emulsifiers—in particular, ethoxylated fatty alcohols—were used for the stabilisation of the PCM droplets/particles. A paraffin with a melting point at 50 °C was used as a nucleating agent that was able to decrease the supercooling. After 100 cooling/heating cycles in a test rig, the DSC measurements of the samples revealed an increased supercooling again and indicated an unstable nucleation agent. Lu and Tassou [12] reported the development of a PCM slurry for cooling applications. A paraffin with a high melting temperature was used as a nucleation agent to reduce the supercooling of the emulsified PCM. The storage stability was tested by storing a sample over eight months and applying 50 heating and cooling cycles. The stability was analysed by comparing the DSC curves before and after the experiment. The emulsion showed no reduction of the effectiveness of the nucleation agent, but separation or creaming of the paraffin droplets was observed, which was, in further experiments, supressed by using thickeners. Zhang et al. [2] dispersed n-octacosane with a melting temperature of 60 °C for use in heat storage applications. As emulsifiers, they used non-ionic surfactant different combinations of Tween and span emulsifiers with a HLB of 12, as well as hydrophobic SiO_2 particles, as nucleating agent to reduce supercooling. The stability was tested first by storing samples for up to six months and, second, by applying five freezing and melting cycles in which the sample was heated to 70 °C and then cooled down again to room temperature. The breaking ratio, which is the ratio between the separated PCM to the whole tested volume, was used to quantify the stability. After the storing experiment, a creaming was observed but no clear trend in the changes of the droplet sizes. In the thermal cycling experiment, separated oil was observed on the samples, and a breaking ratio was determined in a range between 40% and almost 90% for the different emulsions.

Delgado et al. [1] stated that the smaller and more homogenous the particle size distribution is, the more stable a PCM emulsion is for thermal–mechanical cycling and storage time-induced effects. Zhang et al. [13] described the development of PCM emulsions with particle sizes in the nanometre scale using hexadecane as the PCM. The emulsion showed a high stability against phase separation, which was tested under repeated thermal cycles in which the emulsion was cooled below the solidification temperature in an ice bath and heated up again by taking it out of the ice bath and exposing it to room temperature at 25 °C. A high stability was proven by applying this procedure. The DSC measurements revealed supercooling, which was reduced in further experiments by using SiO_2 particles, but nevertheless, the supercooling was still in the range of about 5–10 K.

Vorbeck et al. [14] investigated a PCS based on microencapsulated paraffin with a melting range between 22 and 26 °C and a melting enthalpy of 50 kJ/kg. The PCS had a fraction of 30 wt.% of microcapsules and was tested on stability within a hydraulic test rig in which the PCS was cycled. The test rig contained plate-type heat exchangers and a centrifugal pump. In every cycle, the slurry was cooled below the solidification temperature and heated up above the melting temperature of the PCM again. Nearly 100,000 cycles were applied without observing a significant number of broken microcapsules. The storage capacity was investigated in a storage tank containing 5 m^3 of the slurry, and the results were compared with experiments using water as the storage fluid in the same storage tank. It was shown that it was possible to store 55% more energy than using water, which leads to a 95-min-longer discharge duration. Biedenbach et al. [15] reported the experimental investigation of a PCM emulsion with 35 wt.% n-octadecane in a 500-L storage tank. The storage capacity was compared to water, which was also investigated in the same storage tank. The PCM emulsion showed a supercooling of 5.7 K. Compared to water, the emulsion offered, in the best-case scenario, an improvement of the storage capacity by a factor of 2.6, operating it between 20 and 27 °C, 2.3 in a temperature range between 22 and 29 °C and 1.7 between 15 and 29 °C. The stability of the emulsion was proven by particle size measurements done with samples taken after filling the storage tank after three and after six days of operation. The particle size was decreased and more uniform after three days of moving to slightly bigger sizes after six days. Delgado et al. [16] investigated a PCM emulsion with a low-cost paraffin and solid content between 59% and 61% in a stirred 46-litre tank with a coiled heat exchanger. A 3.5–5.5 times higher heat transfer coefficient was observed when the emulsion was stirred with 290–600 rpm compared to the non-stirred tank. It was found that the stored energy was 80% higher for an operation temperature between 30 and 50 °C and 40% higher when operating it between 30 and 60 °C in comparison to water as the storage fluid. Morimoto et al. [17] investigated the influence of the degree of supercooling on the storage capacity in n-hexadecane and n-octadecane emulsions. A freezing ratio was defined as the relation of the heat stored when the emulsion was cooled to a certain temperature before it was molten again to the total storage capacity of the full-phase transition. Dependent on the cooling temperature, the freezing ratio was in a range between 30% and 60% for emulsions having a mean particle size of 0.25 µm.

Research into PCS development is concentrated on the stabilisation of PCM in emulsions against phase separation (creaming) and on the suppression of supercooling. In many publications, the stability against thermal cycling is tested without applying mechanical stress at the same time. There is little information on how supercooling influences the storage capacity of PCM emulsions in comparison to standard heat transfer fluids. In this study, analyses on the storage capacity for emulsions using different paraffins as PCM in comparison to pure water are presented. All investigated emulsions are prepared using the same formulation, including the nucleation agent. Furthermore, an analysis of the stability is undertaken using repeated thermal cycles and shearing the emulsion with a defined shear rate at the same time (thermal–mechanical stress). The influence of supercooling on the storage capacity is analysed by enthalpy curves obtained via DSC measurements. The degradation due to thermal–mechanical stress is tested by using a rotational rheometer determining the change of viscosity with time, measuring the particle size distribution and the enthalpy via DSC before and after the test.

2. Materials and Methods

2.1. Materials

All presented PCM dispersions were prepared by using linear alcohol ethoxylates from Sasol Germany GmbH, Fritz-Staiger-Straße 15, 25541 Brunsbüttel, German as surfactants. The PCMs, which were emulsified, were Parafol 16-97, Parafol 17-97, Linpar 17, Parafol 18-97, Parafol 19-90 and Parafol 19-97, as well as Parafol 20Z, which were also received from Sasol Germany GmbH. For the paraffins of the Parafol trademark, the first digit stands for the number of carbon atoms in the hydrocarbon chain and the second digit for the purity of the material. For example, Parafol 19-90 and Parafol 19-97 are *n*-alkanes with 19 carbon atoms. These two materials differ solely in their purity, with 90% and 97%, respectively. According to Sasol, Linpar 17 and Parafol 20Z are paraffins with high purities, but there are no exact percentages available. In the following, Parafol is abbreviated to P and Linpar to L. The PCM RT 27 is a blend of paraffins that was manufactured by Rubitherm Technologies GmbH, Imhoffweg 6, 12307 Berlin, Germany with an unknown composition. A second blend was examined using 40 wt.% Parafol 16-97 and 60 wt.% Parafol 18-97, resulting in a melting temperature of 18.7 °C.

2.2. Formulation of the PCM Dispersions

The PCM dispersions were all formulated with 35 wt.% of PCM emulsified in deionised water. The concentration of 35 wt.% of PCM shown in previous internal studies a good compromise between capacity, viscosity and stability. For stabilising the PCM droplets/particles in water, 5 wt.% surfactant was added to the formulation. To decrease the supercooling of the PCM dispersion, a derivate of paraffin wax from Sigma-Aldrich Chemie GmbH, Kappelweg 1, 91625 Schnelldorf, Germany with a melting temperature of 105 °C was added to the PCM as a nucleation agent before dispersing it in water. The mixture of PCM and nucleation agent was heated to 105 °C and stirred for 30 min; then, it was cooled down to 80 °C before emulsifying it in water.

The mixture of Paraffin, nucleation agent, emulsifier and water was dispersed for 3 min using a dispersion machine, magicLAB®, from IKA®-Werke GmbH & Co.KG, Janke & Kunkel-Str. 10, 79219 Staufen, Germany which worked with a rotor–stator principle at a speed of 20,000 rpm. In a following step, the PCM emulsion was homogenised in a high-pressure homogeniser APV 2000 from SPX®, Konrad-Zuse-Straße 25, 47445 Moers, Germany at 200 bar. In total, 200 g of every sample was produced containing 70 g of PCM. The initial weight was done with an accuracy of ±50 mg for PCM and water and with ±5 mg for the emulsifiers and nucleation agent. The maximum weighting error was below ±0.07%, and therefore, this error was neglected for the storage capacity analyses.

2.3. Characterisation

Every PCM dispersion was analysed in terms of its particle size distribution, melting and crystallisation behaviours, enthalpy and stability against thermal–mechanical stress.

The particle size distribution was analysed using a laser diffraction particle analyser, LS13320, from Beckman Coulter GmbH, Europark Fichtenhain B 13, 47807 Krefeld, Germany. The measured values represented the particle diameter. The particle size distribution was measured after production and after the thermal–mechanical stability test.

A DSC (Q200 from TA Instruments, Altendorfstraße 12, 32609 Hüllhorst, Germany) was used to determine the melting/crystallisation temperature and the enthalpy of the fusion/crystallisation. The onset values were chosen as the melting/crystallisation temperature due to its independency of the heating rate. To identify the suitable heating rate for the dispersions, a heating and cooling rate test was performed conformal to the procedure developed in IEA ECES Annex 29/SHC Task 42 [18]. According to the heating rate test, the heating rate was determined to 1 K/min. The measurements were performed in T_{zero}

hermetic aluminium crucibles at the determined heating rate, and the temperature dynamic range varied depending on the melting point of the PCMs. The sample volume was 10 µL. A flow of dry nitrogen was used to purge the measuring cell with 50 mL/h during the measurements. The error of the DSC measurements was given with $\pm 5\%$ for the enthalpy values and ± 0.1 K for the temperature values.

For using PCM emulsions as heat or cold storage fluids, it is important to consider the degree of supercooling, especially when comparing it with water. Supercooling is increasing the necessary temperature range that must be applied to use the full potential of SC. In principle, the larger this temperature range is, the lower is the benefit of PCM systems in comparison to storages using water as the storage fluid. For the determination of SC, a temperature range for charging and discharging is defined as the ΔT_{load}. The SF is given as the volume-specific sensible and latent heat within the temperature range of ΔT_{load} in comparison to the volume-specific sensible heat of water in the same temperature range. For the comparison, a value of 4.19 kJ/(kg·K) as the specific heat capacity for water and a density at 20 °C of 0.998 kg/L were used [19]. As the material datasheets showed no difference in the densities of the used pure paraffins, density measurements were performed using the Parafol 18-97 emulsion. The density was measured 4 times at 35 °C using an Anton Paar DMA 4500 divece from Anton Paar GmbH, Hellmuth-Hirth-Straße 6, 73760 Ostfildern, Germany. It was determined as 0.872 kg/L with a standard deviation of ± 12 kg/L considering all 4 measurements. In this study, different temperature points were chosen to obtain different ΔT_{load} to determine the SC and SF (in the following, called methods a–d):

(a) Temperature range between the start of melting ($T_{m\text{-}start}$) and end of melting ($T_{m\text{-}end}$).
(b) Temperature range of -6 K from the end of melting ($T_{m\text{-}end}$) taking only the melting curve into account.
(c) Temperature range of -6 K from the end of melting ($T_{m\text{-}end}$) considering the crystallisation curve (see Figure 1).
(d) Temperature range between the end of melting ($T_{m\text{-}end}$) and end of crystallisation ($T_{c\text{-}end}$).

$T_{m\text{-}end}$ is determined as the temperature where the difference of enthalpy between the crystallisation and melting curve is smaller than 0.2 kJ/kg the first time when subtracting the melting curve from the crystallisation curve, starting from the onset temperature of melting to higher temperatures. $T_{c\text{-}end}$ is determined in the same way but by decreasing the temperature. $T_{m\text{-}start}$ is chosen visually as the temperature of the melting curve were the slope starts to steepen.

The dependency of the available SC on supercooling is illustrated in Figure 1. The graph shows a DSC measurement of a microencapsulated paraffin that is partly crystallised. The material shows two phase transitions. In the first measuring cycle, the material was cooled down to 20 °C and heated again to 40 °C. In the second cooling, it was only cooled down 25 °C and heated from this temperature again (black, dashed line). As visible in this case, the available SC was determined by the crystallisation curve. In the following analyses, this behaviour was considered to determine the SC based on method c for the different PCM emulsions.

The rotational rheometer MCR 502 from Anton Paar GmbH, Hellmuth-Hirth-Straße 6, 73760 Ostfildern, Germany was used to determine the thermal–mechanical stability of the PCM emulsions. The setup allowed us to apply a defined heating and cooling rate to heat and cool a sample within a defined temperature range, as well as to stress the sample with a defined shear rate at the same time. The stability was tested in a concentric cylinder measuring system, consisting of a measuring body and a measuring cup, wherein the measuring body was rotated. This setup was suitable for low viscous fluids and a relatively large amount of sample could be applied. By using cylindrical geometry, the errors caused by the evaporation of water from the sample were reduced, having a minor effect on the measured results. During the measurements, evaporation was additionally reduced with a solvent trap. The PCM emulsion was heated up and cooled down at a defined rate of 0.05 K/s over 100 thermal cycles while being stressed with a constant shear rate of 100 1/s.

This test will be called the thermal–mechanical stability test in the following. Before and after the thermal–mechanical stability test, a sample of the cycled PCM emulsion was analysed via DSC, and the particle size distribution was measured using the LS13320.

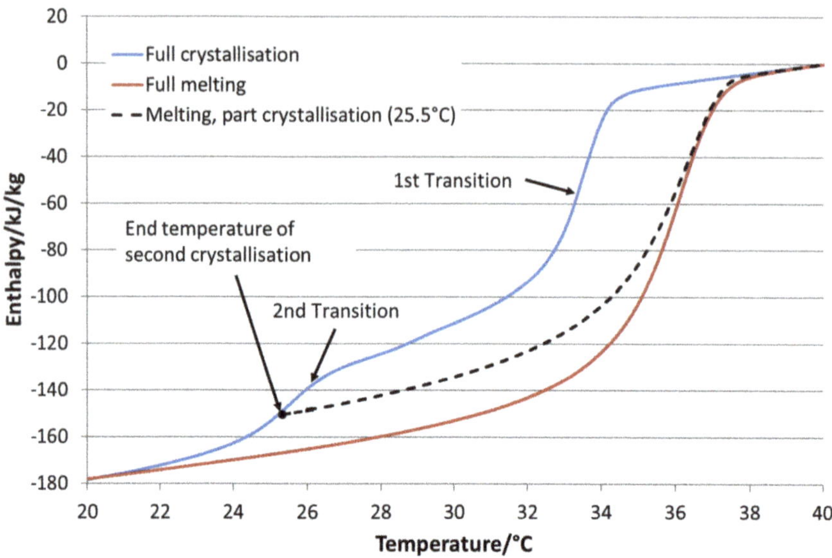

Figure 1. Example for the melting behaviour of a partly crystallised microencapsulated paraffin. The PCM shows two phase transitions in the crystallisation process. The 1st transition is at the higher temperature.

3. Results and Discussion

The used paraffins are characterised via DSC as well. Table 1 summarizes all DSC results and Figure 2 illustrates the variety of phase transitions and enthalpies within the dependency of temperature. All analysed paraffins show supercooling in the form of a hysteresis. Parafol 16-97 (P16-97), Parafol 18-97 (P18-97) and Parafol 20Z (P20Z) are linear *n*-alkanes, with an even number of 16, 18 and 20 carbon atoms. In the DSC, they show a single-phase transition or two-phase transitions (P20Z) that are very close together when crystallising. The phase transitions are sharp and take place in a narrow temperature range. Linpar 17 (L17), Parafol 17-97 (P17-97), Parafol 19-90 (P19-90) and Parafol 19-97 (P19-97) are *n*-alkanes with an odd number of carbon atoms. These *n*-alkanes show a clearly visible second phase transition in both the melting and crystallisation processes.

Table 1. Summary of the most important properties of bulk PCM that are taken for the formulation of PCM dispersions.

PCM/Paraffin	$T_{m\text{-end}}$	$T_{m,\text{ onset}}$	$T_{c,\text{ onset}}$	$T_{m\text{-start\_1st}}$	$T_{m\text{-start-1st}}$	$\Delta H f_{m\text{-tot}}$
	°C	°C	°C	°C	°C	kJ/kg
P16-97	19.6	16.6	15.7	13.9	-	217
P16-97/P18-97	22.3	18.7	21.1	15.4	-	151
L17	24.1	20.5	21.1	17.4	6.1	212
P17-97	24.8	21.2	21.6	18.3	9.3	210
RT27	27.7	24.5	26.1	19.4	-	153
P18-97	30.3	27.3	26.7	23.7	-	223
P19-90	32.4	30.5	30.6	25.9	17.1	207
P19-97	34	31.2	31.5	29	19.2	222
P20Z	36.5	34.3	34.6	29.1	-	211

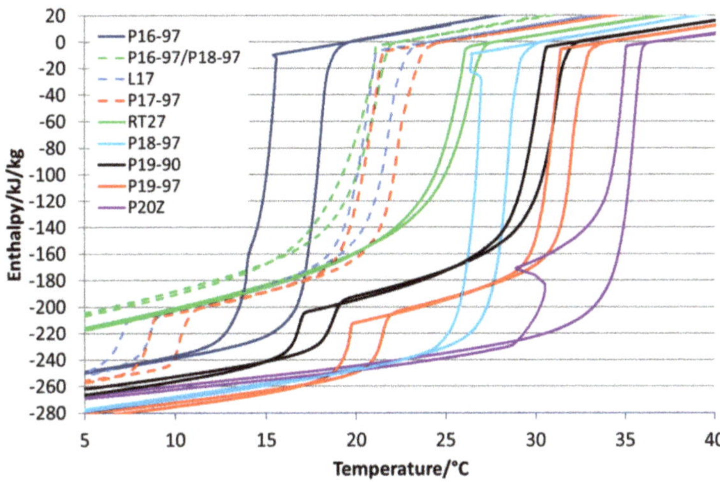

Figure 2. Enthalpy curves of the bulk materials measured with 1 K·min^{-1} via DSC. The shortcuts are given in Table 1.

P17-97 has an onset temperature T_{onset} (melting point) of 21.2 °C, L17 of 20.5 °C, P19-90 of 30.5 °C and P19-97 of 31.2 °C. These paraffins show a smaller hysteresis than most of the pure even-numbered paraffins, but the melting and crystallisation also take place over a broader temperature range. Combined with the two-phase transitions, the temperature difference in an application must then be comparatively large to use the full latent of the storage capacity. For example, P17-97 requires a temperature range of 8 K when only the first phase transition is used and 15 K when the second phase transition should be used as well. The comparison between P19-90 and P19-97 shows how impurities influence the storage capacity and the phase transition temperature. The total storage capacity is reduced from 222 kJ/kg for P19-97 to 207 kJ/kg for P19-90, and the melting temperature (T_{onset}) decreases from 31.2 °C to 30.5 °C.

RT27 is a purchasable blend of *n*-alkanes with paraffin, having a different numbers of carbon atoms in their hydrocarbon chains. The melting point T_{onset} is at 24.5 °C, which is close to that of P18-97 (27.3°C). The phase transition is in a temperature range that is larger than that of P18-97, and the enthalpy for the blend RT27 is 153 kJ/kg, which is 70 kJ/kg less than that of P18-97 (223 kJ/kg).

For the comparison with L17 and P17-97, which are both high-purity odd-numbered paraffins, a blend was mixed using 40 wt.% P16-97 and 60 wt.% P18-97. It has a melting point T_{onset} of 18.7 °C and an enthalpy of 151 kJ/kg. All the investigated blends show a small hysteresis but a broad temperature range for the phase transition in comparison to the pure PCMs.

P20Z is an *n*-alkane with twenty carbon atoms in the hydrocarbon chain. Its melting point T_{onset} is at 34.3 °C, and it offers an enthalpy of 211 kJ/kg. During crystallisation, it also shows two phase transitions, of which the first is the transition to the rotator phase and the second is the transition to the stable triclinic crystal form.

Table 1 summarises the paraffin´s most important properties from the DSC measurements.

3.1. PCM Emulsions

3.1.1. Particle Size Distribution

All PCM are dispersed with 35 wt.% of PCM in water and homogenised. Through homogenisation, a particle size distribution in the sub-micrometre range is reached, and thus, a high stability against creaming is obtained. Figure 3 depicts the particle size distributions for all emulsions. All sizes are below 1 μm. The emulsions with P19-90 and

P20Z have bimodal particle size distributions in the range between 45 and 600 nm, as well as P19-97; all others show a monomodal size distribution in the range between 60 and 500 nm.

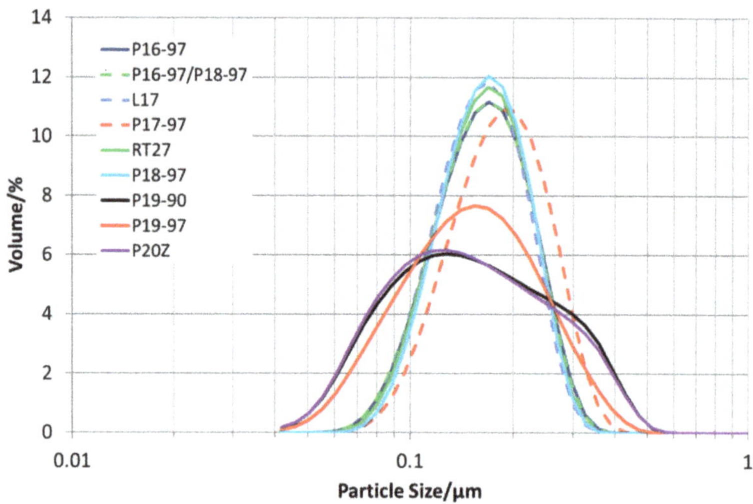

Figure 3. Particle size distribution of PCM dispersions with 35 wt.% of different PCMs dispersed in water.

3.1.2. Storage Capacities

The DSC results are summarised in Table 2 for all the PCM emulsions. The melting enthalpies are in a range between 53 und 61 kJ/kg, which is on the level expected for odd-numbered paraffins due to the melting enthalpy of the first-phase transition of the used bulk paraffins. The even-numbered paraffin emulsions show a melting enthalpy that is less than 35% of the available bulk paraffin´s melting enthalpy (see $Hf_{PCM-1st-35\%}$ in Table 2). This is due to the induction of a second-phase transition through the emulsification process. In comparison to the pure paraffins, all emulsions show a high degree of supercooling ($\Delta T = T_{m\text{-offset}} - T_{c\text{-onset}}$) that is listed in Table 2 and shown in Figure 4 for all the emulsions. Supercooling is reduced with an increasing melting temperature and samples with lower purity. Additionally, those emulsions prepared using paraffins with an odd number of C atoms in their hydrocarbon chain show a lower degree of supercooling than emulsions with even-numbered paraffins.

Table 2. Summary of the main DSC properties for the PCM dispersions, with $\Delta T = T_{m,\,offset} - T_{c,\text{-onset}}$.

Dispersed PCM	$T_{m\text{-end}}$	$T_{m,\,onset}$	$T_{m,\,offset}$	$T_{c,\,onset}$	ΔT	ΔHf_m	ΔHf_c	$\Delta Hf_{PCM\text{-}m\text{-}1st\text{-}35\%}$
	°C	°C	°C	°C	K	kJ/kg	kJ/kg	kJ/kg
P16-97	18.2	16.2	17.9	11.6	6.35	55.6	59.5	75.95
P16-97/P18-97	21.5	18.4	21.0	18.3	2.69	57.9	58.2	52.85
L17	21.7	19.6	21.8	17.5	4.31	60.6	57.8	57.05
P17-97	21.9	20.1	21.9	18.0	3.87	60.6	62.4	55.65
RT27	26.5	23.9	26.4	23.4	2.95	53.1	54	53.55
P18-97	27.6	25.1	27.5	23.0	4.48	58.2	62.5	78.05
P19-90	32.0	29.0	31.6	28.8	2.8	60.17	60.93	54.25
P19-97	32.4	29.9	32.0	28.8	3.22	59.3	62.3	58.45
P20Z	36.0	31.4	35.9	32.9	3	59	59.1	73.85

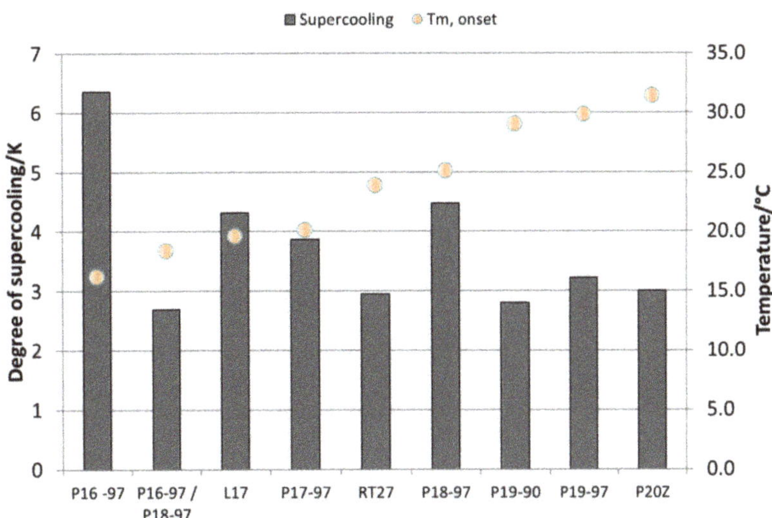

Figure 4. Degree of supercooling and onset temperature of melting for the different emulsions.

Figure 5 depicts the enthalpy curves for all PCM emulsions dependent on the temperature. The graph reveals that all emulsions have at least two-phase transitions in the measured temperature range, except the one prepared with a mixture of P16-97 and P18-97. It might have a second transition below the measured temperature range, but this could not be proved, as it was not possible to cool further down due to the freezing of the emulsion´s water fraction. The curves also show that those emulsions prepared with mixtures of paraffins (P16-97/P18-97, RT27) show broader melting and crystallisation ranges than those using purer paraffins. Figure 6 compares the enthalpy curves of L17, P17-97 and the mixture of P16-97/P18-97 emulsions. It is visible that L17 and the P16-97/P18-97 mixture offer almost the same enthalpy differences in the main transition range, whereas P17-97 melts at a higher temperature and shows a higher change in the enthalpy. Both the bulk PCM L17 and P17-97 have around 210 kJ/kg, while the blend of P16-97/P18-97 is 28% lower with 151 kJ/kg. After emulsification, the enthalpy of this mixture is only about 4.5% lower than that of the L17 and P17-97 emulsions (57.9 k/kg versus 60.6 kJ/kg). A similar effect can be seen for the P18-97 and RT27 emulsions. While the enthalpy of bulk RT-27 is 31% lower, it is only 14% lower when it is emulsified.

Despite the melting enthalpy for the bulk P16-97 being slightly higher (217 kJ/kg) than that of P20Z (211 kJ/kg), emulsified P20Z shows, with 59.0 kJ/kg, a slightly higher melting enthalpy than the emulsified P16-97, which provides 55.6 kJ/kg. One reason for this is the higher share of enthalpy within the second phase transition at a lower temperature for the P16-97 emulsion, which has 12.2 kJ/kg versus 8.6/kJ for the P20Z emulsion.

The storage capacities and temperature ranges for the emulsions, calculated according to methods a–d, are shown in Figure 7 and listed in for method a and b in Table 3 and and for method c and d in Table 4. As expected, the larger the chosen temperature range, the more storage capacity is available, except for the analyses considering a 6 K temperature difference and taking the crystallisation curves into account according to method c. These analyses revealed a much lower storage capacity, as expected from the enthalpy change of the melting curve. Here, those materials with a high degree of supercooling show the lowest storage capacity. The comparisons of the storage capacities with those that would be obtained using water as the storage medium are shown in Figure 8. Here, a massive reduction of the benefits due to the supercooling phenomena and due to broad-phase transition temperature ranges is visible. The analyses that take only the melting curves into account (method a) show the maximum possible SF operating a system at the phase transition from a solid to liquid without any supercooling. Considering only the melting curve with a temperature range of 6 K (according to method b), the benefit is reduced for all materials but shows the realistic maximum benefit compared to water as the storage material in a real storage application. With the given supercooling and melting, as well as crystallisation ranges, P20Z, P19-97 and P17-97 perform best with a SF around 2.5. Considering the 6-K limit according to method c, RT27 offers a 1.8 higher SF than P18-97. Using the full temperature range (method d), P18-97 shows with 2.22 versus 2.06 (RT27) a 7.8% higher SF.

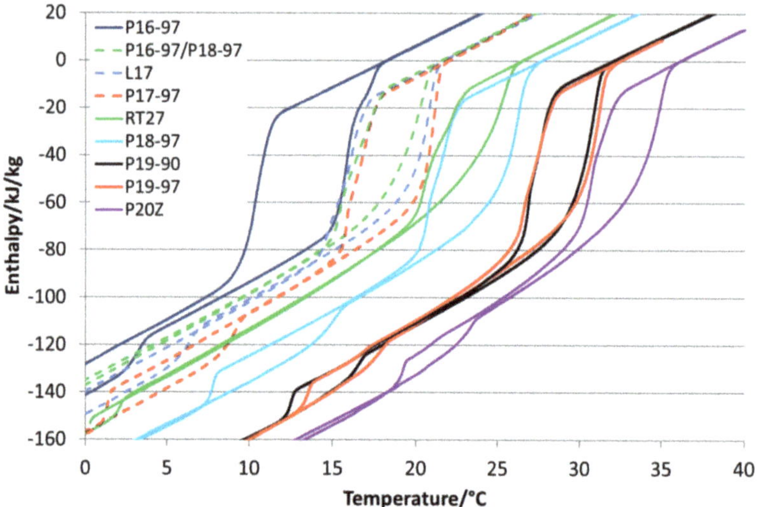

Figure 5. Enthalpy curves of the PCM emulsions with a PCM fraction of 35 wt.% each. For each curve, the enthalpy is set to 0 at $T_{m\text{-end}}$.

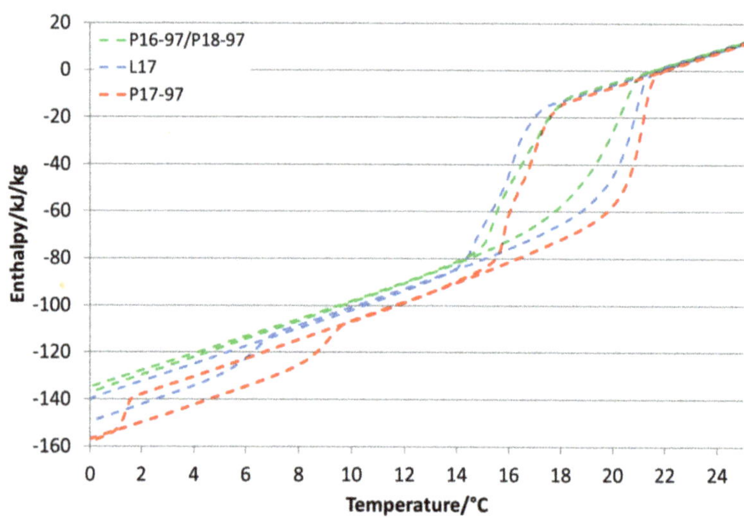

Figure 6. Enthalpy curves for the P16-97/P18-97, L17 and P17-97 emulsions.

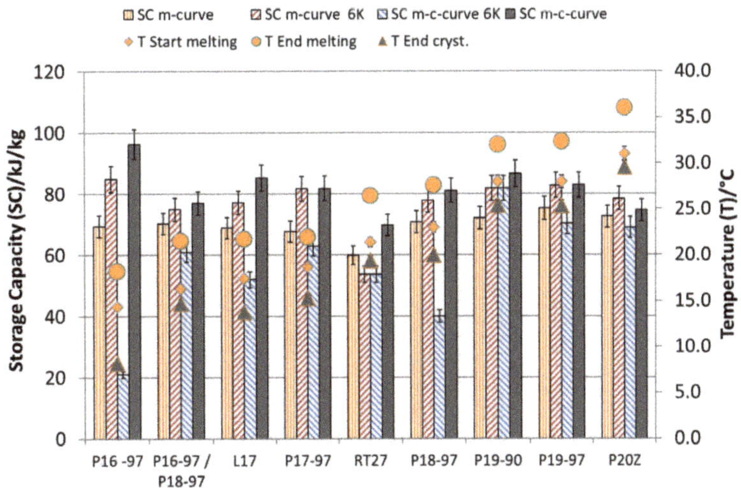

Figure 7. Storage capacities according to the different temperature ranges: m-curve (method a), m-curve 6 k (method b), m-c-curve (method c) and m-c curve (method d) (see Section 2.3 Characterisation. Storage factors and temperature differences for the melting process and to acquire the total storage capacity for the first phase-transition m-curve (method a), m-curve 6 k (b), m-c-curve (c) and m-c curve (d) (see Section 2.3 Characterisation).

Table 3. Data for the different emulsions for the analyses according to methods a and b.

Sample	Based on the Melting Curve (a)						Based on the Melting Curve, dT 6K (b)			
	$T_{m\text{-}end}$	$H_{m\text{-}end}$	$T_{m\text{-}start}$	ΔT_{load}	$\Delta_{Hm\text{-}end\text{-}start}$	SF	$T_{m\text{-}end\text{-}6K}$	ΔT_{load}	$\Delta H_{m\text{-}end\text{-}6K}$	SF
	°C	kJ/kg	°C	K	kJ/kg	-	°C	K	kJ/kg	-
P16-97	18.2	0.0	14.4	3.84	69.37	3.77	12.2	6	84.84	2.95
P16-97/P18-97	21.5	0.0	16.4	5.14	70.18	2.85	15.5	6	74.87	2.60
L17	21.7	0.0	17.5	4.27	68.82	3.36	15.7	6	77.1	2.68
P17-97	21.9	0.0	18.7	3.2	67.64	4.41	15.9	6	81.71	2.84
RT27	26.5	0.0	21.4	5.09	59.79	2.45	20.5	6	53.67	1.87
P18-97	27.6	0.0	23.0	4.63	70.79	3.19	21.6	6	77.81	2.70
P19-90	32.0	0.0	28.0	4.01	72.04	3.75	26.0	6	81.83	2.84
P19-97	32.4	0.0	28.0	4.35	75.22	3.61	26.4	6	82.8	2.88
P20Z	36.0	0.0	31.0	5.02	72.5	3.01	30.0	6	78.43	2.73

Table 4. Data for the different emulsions for the analyses according to methods c and d.

Sample	Based on the Melting and Cryst. Curve, dT 6K (c)				Based on the End of Melting and End of Cryst. Curve (d)			
	$T_{c\text{-}m\text{-}6K}$	ΔT_{load}	$\Delta H_{c\text{-}m\text{-}6K}$	SF	$T_{c\text{-}end}$	ΔT_{load}	$\Delta H_{m\text{-}end\text{-}c\text{-}end}$	SF
	°C	K	kJ/kg	-	°C	K	kJ/kg	-
P16-97	12.2	6	20.9	0.73	8.2	10.00	96.29	2.01
P16-97/P18-97	15.5	6	60.8	2.11	14.7	6.77	76.92	2.37
L17	15.7	6	51.96	1.81	13.8	7.88	85.18	2.25
P17-97	15.9	6	62.88	2.19	15.3	6.62	81.79	2.58
RT27	20.5	6	53.66	1.87	19.4	7.06	69.72	2.06
P18-97	21.6	6	40.03	1.39	20.0	7.62	81.06	2.22
P19-90	26.0	6	81.83	2.84	25.4	6.60	86.51	2.73
P19-97	26.4	6	70.13	2.44	25.3	7.02	82.82	2.46
P20Z	30.0	6	68.95	2.40	29.6	6.46	74.57	2.41

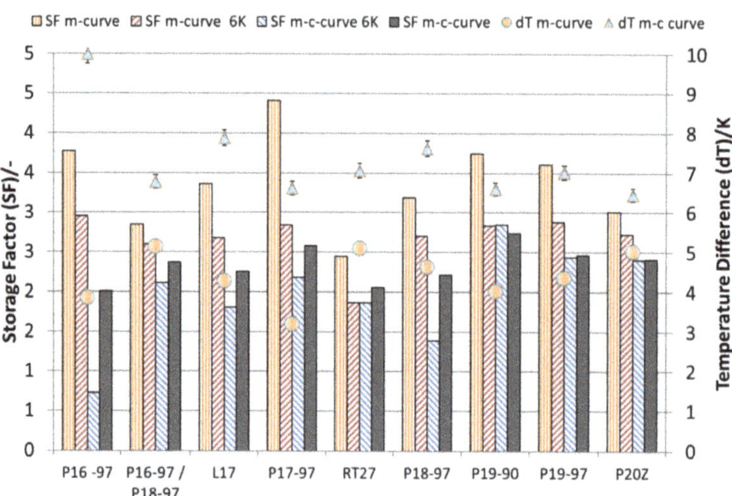

Figure 8. Storage factors and temperature differences for the melting process and to acquire the total storage capacity for the first phase-transition m-curve (method a), m-curve 6 k (b), m-c-curve (c) and m-c curve (d) (see Section 2.3 Characterisation).

3.1.3. Thermal–Mechanical Stability

P16-97, L17, P18-97, P19-90 and P20Z emulsions are chosen for the stability evaluations. All tested PCM emulsions passed this stability test over 100 thermal–mechanical cycles without optically visible changes like phase separation or creaming. Figures 9–13 depict the results for the measured emulsions. The rheological data show different changes in the viscosity over the 100 cycles, which come with the changes in the particle size distribution. In the viscosity measurements, the lower viscosity value indicates the emulsion´s viscosities at 35 °C and the upper value those at 5 °C, respectively. At the beginning of the tests, the emulsions show viscosities between 20 and 50 mPa·s, except P19-90, which ranges between 25 and almost 95 mPa·s. Looking at the development of viscosity over the 100 cycles, the samples exhibit quite different changes. For P16-97, the viscosity at 35 °C is kept more or less constant at 20 mPa·s, while the upper value at 5 °C rises very linearly with every cycle, up to 80 mPa·s after 100 cycles. This comes with a change in the particle size distribution, from a monomodal distribution at the beginning to a bimodal distribution with a fraction of smaller sizes and a larger fraction at bigger sizes. L17 shows a rise in viscosity at 35 °C from 20 to 32 mPa·s and from almost 55 mPa·s to 80 mPa·s after 100 cycles at 5 °C. The particle size increase is minor, and the peak is slightly narrower than at the beginning. P18-97 keeps the lower viscosity at 20 mPa·s, and the upper values are also more or less constant with higher values for single peaks. No significant change in the size distribution is observed. The PCM emulsion P19-90 shows the highest viscosity values in this test. It reaches over 120 mPa·s after 100 cycles. The size distribution also becomes bimodal, but contrary to P16-97, the share that turned to smaller sizes is larger than the one grown to bigger sizes. P20Z also shows a viscosity between 20 and 55 mPa·s at the beginning of the cycle test. While cycled, both values rise, but the upper value at 5 °C rises faster than the lower value at 35 °C.

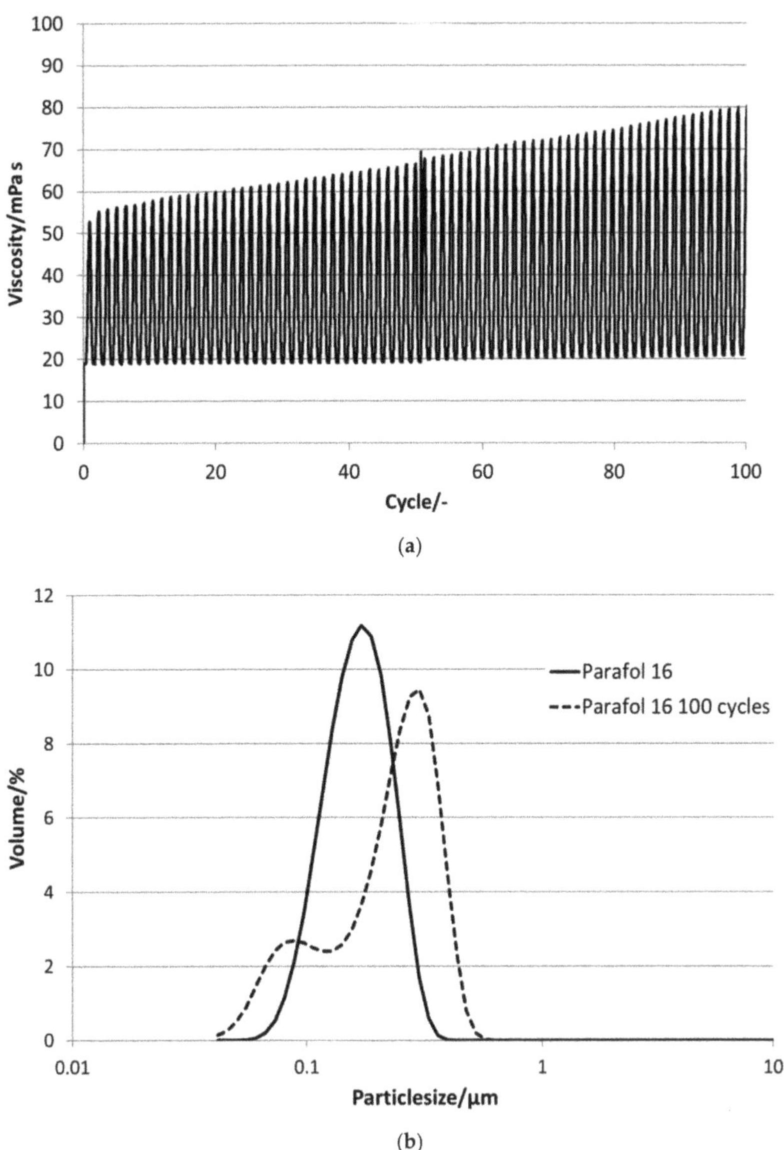

Figure 9. Thermal–mechanical stability test of Parafol 16-97: (**a**) viscosity measured between 5 and 35 °C over 100 cycles, and (**b**) particle size distribution before and after 100 cycles.

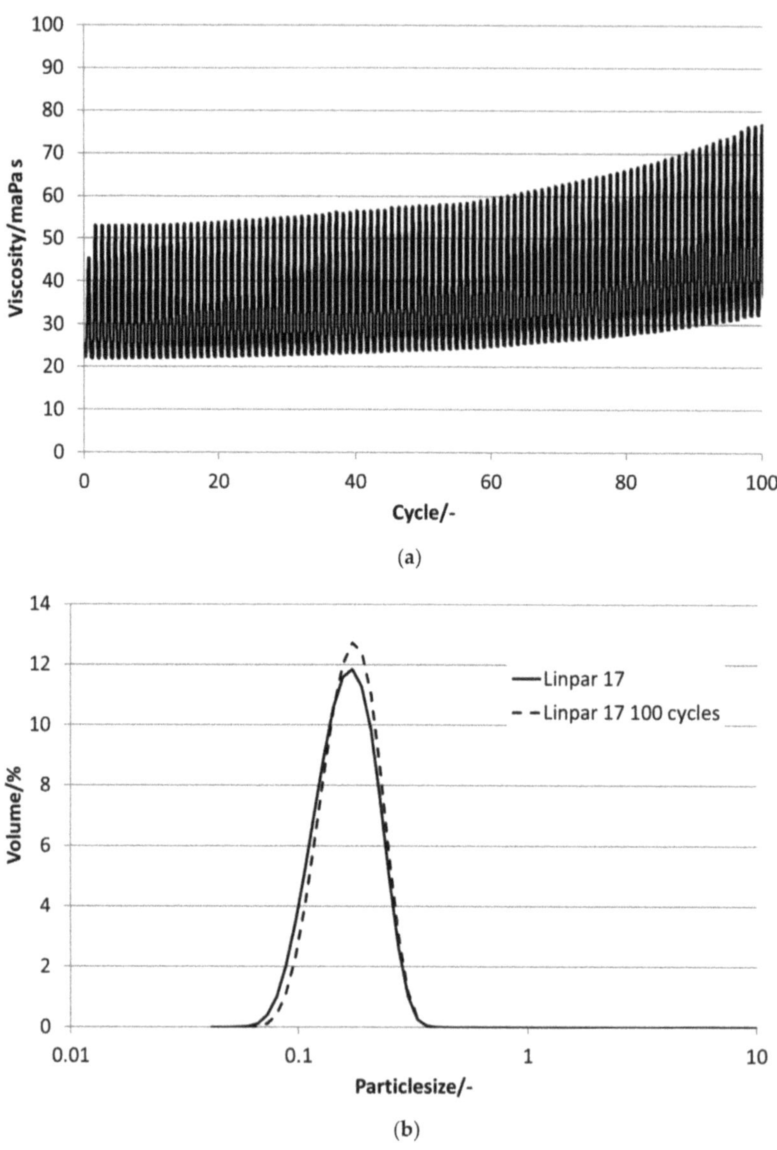

Figure 10. Thermal–mechanical stability test of Linpar 17: (**a**) viscosity between 5 and 35 °C over 100 cycles, and (**b**) particle size distribution before and after 100 cycles.

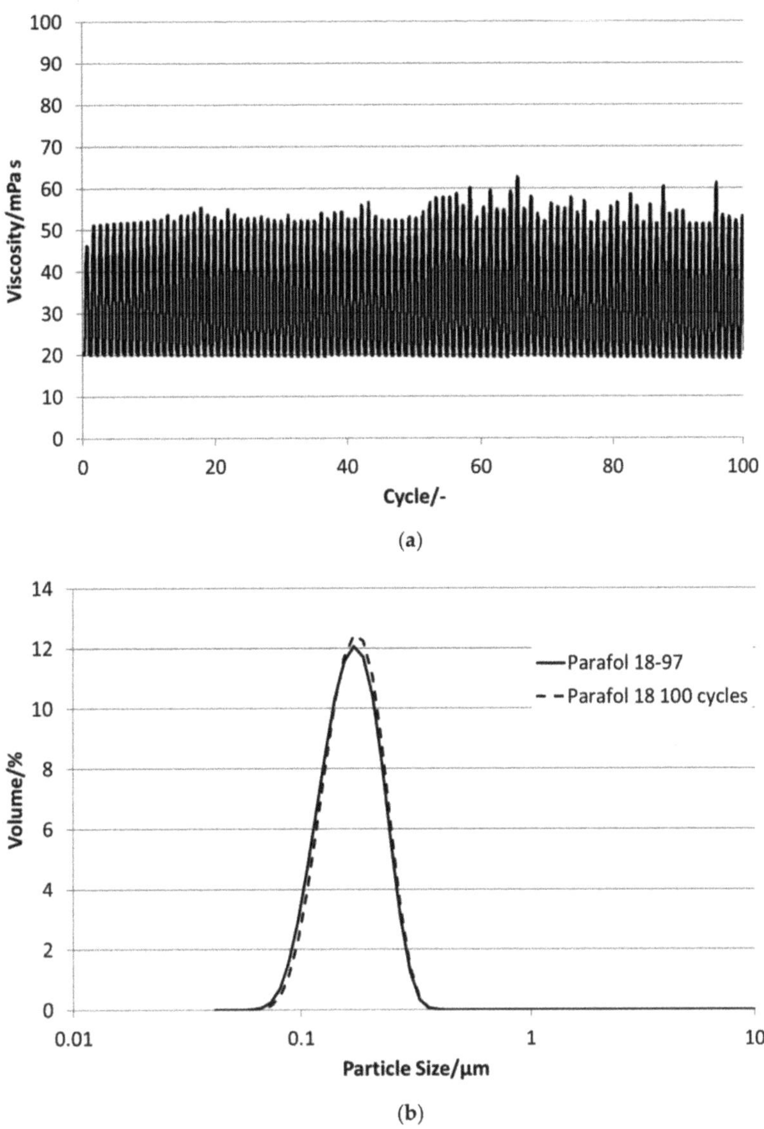

Figure 11. Thermal–mechanical stability test of Parafol 18-97: (**a**) viscosity between 5 and 35 °C over 100 cycles, and (**b**) particle size distribution before and after 100 cycles.

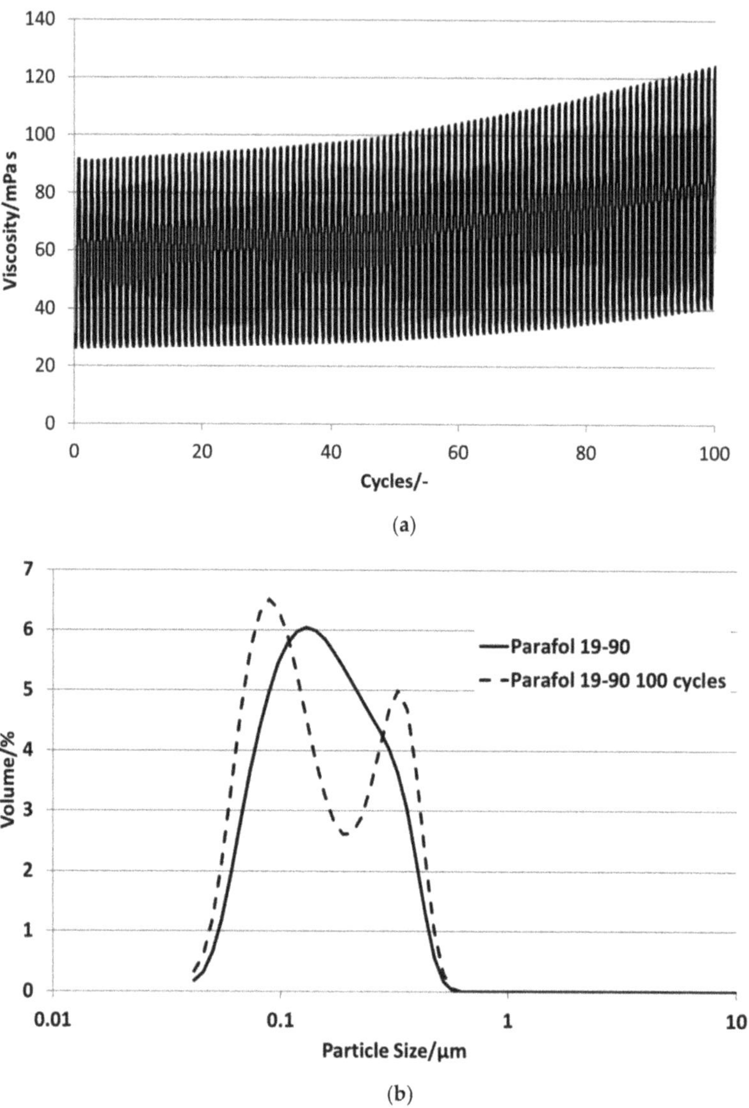

Figure 12. Thermal–mechanical stability test of Parafol 19-90: (**a**) viscosity between 5 and 35 °C over 100 cycles, and (**b**) particle size distribution before and after 100 cycles.

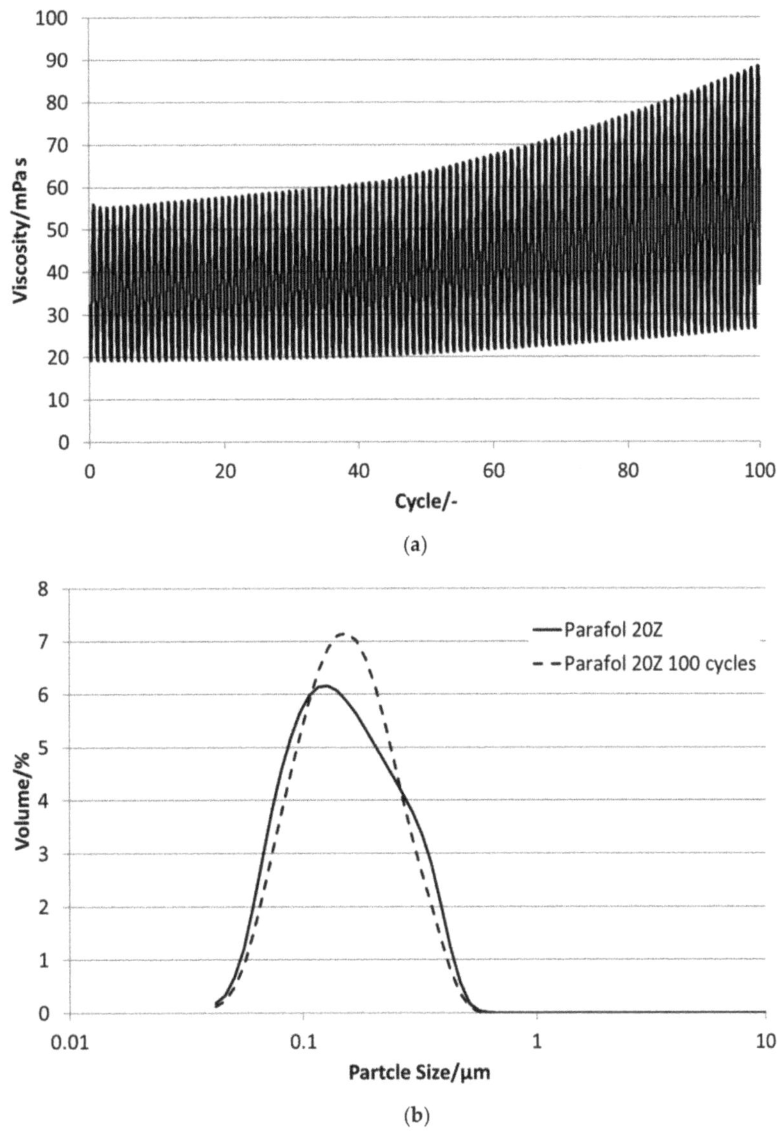

Figure 13. Thermal–mechanical stability test of Parafol 20z: (**a**) viscosity between 5 and 35 °C over 100 cycles, and (**b**) particle size distribution before and after 100 cycles.

Figure 14 shows the comparison of enthalpy curves determined before and after this test. It is visible that all samples show minor changes in the crystallisation curve except P18-97. P16-97, L17 and P19-90 show slightly but not significantly deeper solidification temperatures after the test that might be an indication of an unstable nucleation seed. P20Z reveals an increase in the phase transition temperatures from a liquid to solid, which could be induced by an increase in particle size. Therefore, in general, it can be stated that those emulsions showing minor changes in the viscosity and in the particle size distribution also revealed minor changes in the phase transition behaviour.

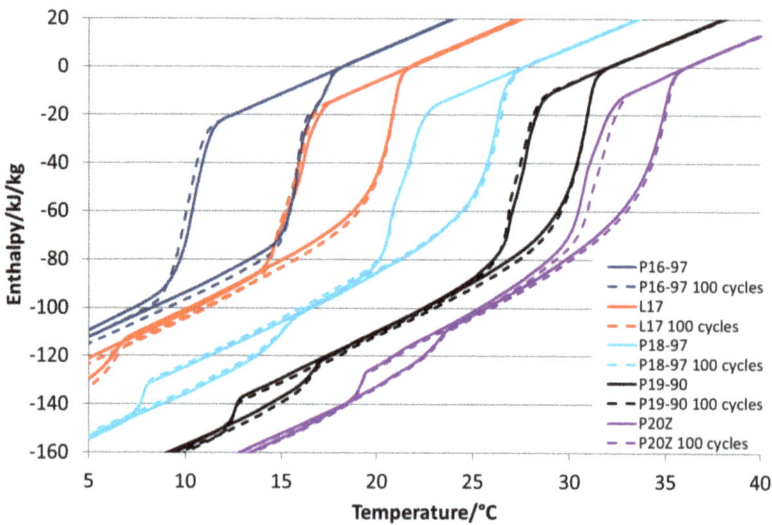

Figure 14. Comparison of the melting and crystallisation behaviours before and after the thermal-mechanical stability test.

4. Conclusions

In this study, even- and odd-numbered paraffins were used to prepare PCM emulsions and mixtures. The DSC analyses of the paraffins showed the two-phase transitions of the odd-numbered paraffins. The even-numbered paraffins shown in the DSC were only one-phase transitions, except for P20Z, which was a two-phase transition that was close together in the measurements. Furthermore, all used paraffin mixtures showed less enthalpy than the pure materials and a larger temperature range for the phase transition from a solid to liquid and, also, less supercooling. While all used pure paraffins offered total enthalpies in the range between 207 and 223 kJ/kg for the phase transition from a solid to liquid, the enthalpies of the mixtures P16-97/P18-97 and RT27 were 151 kJ/kg and 157 kJ/kg, respectively. This was on the same level with the enthalpies of the first transition of the measured odd-numbered paraffins, which were about 75–77% of their total enthalpy.

The supercooling and storage factors of the PCM emulsions were influenced by the emulsified PCM and nucleation agent. The odd- and even-numbered paraffins showed different crystallisations in the bulk but similar behaviours when emulsified. Odd-numbered paraffins kept the two-phase transitions of the bulk materials, while the investigated even-numbered paraffins showed a second phase transition (different from the bulk materials) that indicated a phase transition from the liquid to a rotator phase (first transition) and from the rotator phase to the stable phase (second transition). This behaviour led to a reduction of up to 27% of the bulk paraffin´s phase transition enthalpy for the first transition, as the emulsion P16-97, for example, showed. It was shown that the purer materials exhibited more supercooling when emulsified than the blends or mixtures. The results revealed that, due to this effect, the mixtures and blends were better in terms of the SF when small temperature intervals were given, which was shown by using a temperature difference of 6 K (comparing RT27 with P18-97 and P19-90 with P19-97). Additionally, the mixture of P16-97 and P18-97 had a SF within a 6-K range that was on the same level as P17-97 (2.11 versus 2.19). If a large melting enthalpy of a bulk material suggests a large SF for a PCM emulsion, these examples showed that this could not be transferred to the PCM emulsion directly, because the supercooling and limited temperature intervals of the applications needed to be taken into account and strongly influenced the SF and, therefore, the advantage of the emulsions over conventional heat transfer fluids.

One aspect of the analysed PCMs that could be transferred to the PCM emulsions was its tendency of supercooling with a decreasing melting temperature. The shorter the PCM´s hydrocarbon chain, the larger the degree of supercooling. Furthermore, the supercooling of the PCM emulsions was higher using purer materials. This effect was shown for Parafol 19-90 and Parafol 19-97, as well as for RT27 and P18-97 emulsions.

Therefore, in terms of costs, it is favourable to have knowledge about the specific temperature interval of an application to consider whether it is more cost-efficient to use not-so-pure but cheaper PCMs providing a good storage capacity within small temperature intervals or to choose the more expensive, purer PCMs to provide more storage capacity when the usable temperature interval is a bit broader.

All the investigated PCM emulsions had similar particle sizes. This indicated that, for the investigated emulsions, the particle size distribution was dependent on the formulation rather than on the used PCM. The thermal–physical properties were measured before and after cycling and showed comparable results for P16-97, L17, P18-97, P19-90 and P20Z emulsions. None of them showed visible instabilities, such as creaming or phase separation, after 100 thermal–mechanical cycles in the rheometer, but a major or minor change in the behaviours and properties was visible for the measured viscosities, particle size distributions and crystallisation behaviours. Especially, the changes in the viscosity and particle size were significant and revealed differences. In comparison, the changes in the melting and solidification behaviours were not so pronounced, and it might be more difficult to draw conclusions about stability from these DSC results. Nevertheless, there were some changes visible that might indicate, e.g., the stability of the crystallisation seed and, due to decreased supercooling, the influence of a change in the particle sizes. In general, all used measuring techniques could reveal a change of properties and, therefore, instability before creaming or phase separation was visible and, therefore, provided an indication of the long-term stability or instability of the PCM emulsions. So far, it is not known how the observed changes affect the stability of these PCM emulsions under real usage or what they mean in terms of useful life applications. Therefore, in future works, the changes of the material properties due to the thermal–mechanical stability test undertaken in the rheometer have to be compared with the stabilities obtained under real applicational conditions, with the aim of developing a model to predict the real stability of PCM emulsions on the basis of this accelerated lab test.

Author Contributions: Conceptualization, S.G. (Stefan Gschwander) and S.N.; methodology, S.G. (Stefan Gschwander) and S.N.; investigation, S.G. (Stefan Gschwander) and S.N.; resources, S.G. (Stefan Gschwander); data curation, S.G. (Stefan Gschwander) and S.N.; writing—original draft preparation, S.G. (Stefan Gschwander) and S.N.; writing—review and editing, S.G. (Sebastian Gamisch), M.K., F.K. and T.H.; visualization, S.G. (Stefan Gschwander) and S.N.; supervision, S.G. (Stefan Gschwander) and T.H.; project administration, S.G. (Stefan Gschwander); funding acquisition, S.G. (Stefan Gschwander). All authors have read and agreed to the published version of the manuscript.

Funding: The research was funded by the German Federal Ministry of Economics and Energy via several research projects (grant no.: 03ESP357).

Institutional Review Board Statement: Excluded.

Informed Consent Statement: Informed consent was obtained from all subjects involved in the study.

Data Availability Statement: The data presented in this study are available on request from the corresponding author. The data are not publicly available due to restriction given by partners.

Acknowledgments: The authors are grateful to the Project Management Jülich for the administrative support.

Conflicts of Interest: No conflicts of Interest.

Nomenclature

DSC	**Differential Scanning Calorimeter**
HLB	Hydrophilic lipophilic balance
H_f	heat of fusion
$H_{f\text{-}1st}$	heat of fusion first transition
$H_{f\text{-}2nd}$	heat of fusion second transition
$H_{f\text{-}PCM\text{-}1st\text{-}35\%}$	35% of the heat of fusion of the PCM's first transition
$H_{f\text{-}tot}$	heat of fusion of all transitions
$DH_{m\text{-}curve}$	enthalpy difference of melting curve including sensible heat
$DH_{m\text{-}end\text{-}start}$	enthalpy difference between end and start of melting curve
$DH_{m\text{-}end\text{-}6K}$	enthalpy difference—6 K from end of melting, based on melting curve
$DH_{c\text{-}m\text{-}end\text{-}6K}$	enthalpy difference—6 K from end of melting, based on crystallisation curve
$DH_{m\text{-}end\text{-}c\text{-}end}$	enthalpy difference based on end of melting and on end of crystallisation
L	Linpar
P	Parafol
PCM	**Phase-Change Material**
PCS	Phase-change slurry
SC	storage capacity
SF	storage factor
T_{onset}	onset temperature
$T_{m\text{-}star}$	start temperature of melting process
$T_{m\text{-}end}$	end temperature of the melting process
$T_{m\text{-}onset}$	onset temperature of the melting peak
$T_{c\text{-}onset}$	onset of the crystallisation curve
$T_{m\text{-}start\text{-}1st}$	start melting first transition (main peak)
$T_{m\text{-}start\text{-}2nd}$	start melting second transition (at lower temperature)
$T_{m\text{-}end\text{-}6k}$	Temperature—6 K from end of melting
DT_{load}	temperature difference based on different temperatures

References

1. Delgado, M.; Lázaro, A.; Mazo, J.; Zalba, B. Review on phase change material emulsions and microencapsulated phase change material slurries: Materials, heat transfer studies and applications. *Renew. Sustain. Energy Rev.* **2012**, *16*, 253–273. [CrossRef]
2. Zhang, X.; Wu, J.; Niu, J. PCM-in-water emulsion for solar thermal applications: The effects of emulsifiers and emulsification conditions on thermal performance, stability and rheology characteristics. *Sol. Energy Mater. Sol. Cells* **2016**, *147*, 211–224. [CrossRef]
3. El Rhafiki, T.; Kousksou, T.; Jamil, A.; Jegadheeswaran, S.; Pohekar, S.D.; Zeraouli, Y. Crystallization of PCMs inside an emulsion: Supercooling phenomenon. *Sol. Energy Mater. Sol. Cells* **2011**, *95*, 2588–2597. [CrossRef]
4. Youssef, Z.; Delahaye, A.; Huang, L.; Trinquet, F.; Fournaison, L.; Pollerberg, C.; Doetsch, C. State of the art on phase change material slurries. *Energy Convers. Manag.* **2013**, *65*, 120–132. [CrossRef]
5. Huang, L.; Doetsch, C.; Pollerberg, C. Low temperature paraffin phase change emulsions. *Int. J. Refrig.* **2010**, *33*, 1583–1589. [CrossRef]
6. Inaba, H.; Morita, S.-I. Cold heat-release characteristics of phase-change emulsion by air-emulsion direct-contact heat exchange method. *Int. J. Heat Mass Transf.* **1996**, *39*, 1797–1803. [CrossRef]
7. He, B.; Setterwall, F. Technical grade paraffin waxes as phase change materials for cool thermal storage and cool storage systems capital cost estimation. *Energy Convers. Manag.* **2002**, *43*, 1709–1723. [CrossRef]
8. Montenegro, R.; Landfester, K. Metastable and stable morphologies during crystallization of alkanes in miniemulsion droplets. *Langmuir* **2003**, *19*, 5996–6003. [CrossRef]
9. Hagelstein, G.; Gschwander, S. Reduction of supercooling in paraffin phase change slurry by polyvinyl alcohol. *Int. J. Refrig.* **2017**, *84*, 67–75. [CrossRef]
10. Günther, E.; Huang, L.; Mehling, H.; Dötsch, C. Subcooling in PCM emulsions—Part 2: Interpretation in terms of nucleation theory. *Thermochim. Acta* **2011**, *522*, 199–204. [CrossRef]
11. Zhang, S.; Niu, J. Experimental investigation of effects of supercooling on microencapsulated phase-change material (MPCM) slurry thermal storage capacities. *Sol. Energy Mater. Sol. Cells* **2010**, *94*, 1038–1048. [CrossRef]
12. Lu, W.; Tassou, S.A. Experimental study of the thermal characteristics of phase change slurries for active cooling. *Appl. Energy* **2012**, *91*, 366–374. [CrossRef]
13. Zhang, X.; Niu, J.; Wu, J. Evaluation and manipulation of the key emulsification factors toward highly stable PCM-water nano-emulsions for thermal energy storage. *Sol. Energy Mater. Sol. Cells* **2021**, *219*, 110820. [CrossRef]

14. Vorbeck, L.; Gschwander, S.; Thiel, P.; Lüdemann, B.; Schossig, P. Pilot application of phase change slurry in a 5m3 storage. *Appl. Energy* **2013**, *109*, 538–543. [CrossRef]
15. Biedenbach, M.; Poetzsch, L.; Gschwander, S. Characterization of an n-octadecane PCS in a 0.5 m^3 storage tank test facility. *Int. J. Refrig.* **2019**, *104*, 76–83. [CrossRef]
16. Delgado, M.; Lázaro, A.; Mazo, J.; Peñalosa, C.; Marín, J.M.; Zalba, B. Experimental analysis of a coiled stirred tank containing a low cost PCM emulsion as a thermal energy storage system. *Energy* **2017**, *138*, 590–601. [CrossRef]
17. Morimoto, T.; Kawana, Y.; Saegusa, K.; Kumano, H. Supercooling characteristics of phase change material particles within phase change emulsions. *Int. J. Refrig.* **2019**, *99*, 1–7. [CrossRef]
18. Task 42/Annex 29 Thermal Material Database: T4229 DSC-Standard PCM version 03/03/2015. Available online: https://thermalmaterials.org/measurement-standards (accessed on 2 December 2020).
19. Glück, B. *Zustands- und Stoffwerte: Wasser, Dampf, Luft; Verbrennungsrechnung*; Verl. für Bauwesen: Berlin, Germany, 1991; ISBN 3-345-00487-9.

Article

Experimental Study on Two PCM Macro-Encapsulation Designs in a Thermal Energy Storage Tank

David Vérez, Emiliano Borri, Alicia Crespo, Boniface Dominick Mselle, Álvaro de Gracia, Gabriel Zsembinszki and Luisa F. Cabeza *

GREiA Research Group, Universitat de Lleida, Pere de Cabrera s/n, 25001 Lleida, Spain; david.verez@udl.cat (D.V.); emiliano.borri@udl.cat (E.B.); alicia.crespo@udl.cat (A.C.); boniface.mselle@udl.cat (B.D.M.); adegracia@diei.udl.cat (Á.d.G.); gabriel.zsembinszki@udl.cat (G.Z.)
* Correspondence: luisaf.cabeza@udl.cat; Tel.: +34-973-003-576

Abstract: The use of latent heat thermal energy storage is an effective way to increase the efficiency of energy systems due to its high energy density compared with sensible heat storage systems. The design of the storage material encapsulation is one of the key parameters that critically affect the heat transfer in charging/discharging of the storage system. To fill the gap found in the literature, this paper experimentally investigates the effect of the macro-encapsulation design on the performance of a lab-scale thermal energy storage tank. Two rectangular slabs with the same length and width but different thickness (35 mm and 17 mm) filled with commercial phase change material were used. The results show that using thinner slabs achieved a higher power, leading to a reduction in the charging and discharging time of 14% and 30%, respectively, compared with the thicker slabs. Moreover, the variation of the heat transfer fluid flow rate has a deeper impact on the temperature distribution and the energy charged/released when thicker slabs were used. The macro-encapsulation design did not have a significant impact on the discharging efficiency of the tank, which was around 85% for the operating thresholds considered in this study.

Keywords: thermal energy storage; latent heat thermal energy storage; phase change materials (PCM); macro-encapsulation; rectangular slab; experimental study

Citation: Vérez, D.; Borri, E.; Crespo, A.; Mselle, B.D.; de Gracia, Á.; Zsembinszki, G.; Cabeza, L.F. Experimental Study on Two PCM Macro-Encapsulation Designs in a Thermal Energy Storage Tank. *Appl. Sci.* **2021**, *11*, 6171. https://doi.org/10.3390/app11136171

Academic Editor: Ioannis Kartsonakis

Received: 25 May 2021
Accepted: 30 June 2021
Published: 2 July 2021

Publisher's Note: MDPI stays neutral with regard to jurisdictional claims in published maps and institutional affiliations.

1. Introduction

The use of thermal energy storage (TES) has been proved as an effective way to enhance the penetration of renewable energy into energy systems. Amongst all thermal storage technologies, latent heat thermal energy storage (LHTES) received the attention of several researchers over the last decade due to its high energy density and the wide range of applications [1]. Buildings, for example, represent one of the most common applications of the integration of LTHES as an active or passive system [2–4]. For the active systems, TES can be used in HVAC components or systems to balance the supply of domestic hot water and heating/cooling demand when renewables are used [5,6], or to reduce the energy consumption through peak load shifting [7], or free cooling techniques [8]. On the other hand, passive systems are directly integrated into the building envelope to reduce the energy demand [9,10]. Other common applications where LTHES can be integrated include solar thermal power plants, such as concentrated solar power (CSP) [11], solar cooling applications [12], district heating or cooling [13], waste heat recovery [14], solar process heat [15], or cryogenic applications [16].

The principle behind LHTES is the use of phase change materials (PCM) as the storage medium, allowing to store thermal energy at a nearly constant temperature exploiting the latent heat during the phase transition, for which the most common one is from solid to liquid to minimize the impact of volume expansions [17]. One of the weaknesses of PCM is its low thermal conductivity that negatively affects the thermal power involved in the charging and discharging processes of the energy storage system. Indeed, this represents one of the main challenges facing the implementation of PCM in various applications. However, different strategies and techniques that can be used to improve thermal conductivity were investigated in the literature. The main solutions that were extensively studied are the increase in the convection coefficient of heat transfer by means of dynamic systems, the addition of particles (such as carbon elements, metallic particles, and nanoparticles), the inclusion of PCM in a metallic matrix, and the increase in the heat transfer area by using fins, and micro and macro-encapsulation [18–20].

On one hand, PCM micro-encapsulation allows increasing heat transfer surface between the PCM and the heat transfer fluid. However, for PCM microencapsulation, complex and expensive processes are needed, such as spray drying (physical method) or interfacial polymerization (chemical methods) [21]. On the other hand, macro-encapsulation requires a simpler making process resulting in a lower cost [22]. Furthermore, larger sizes of the container also allow an increase in the mechanical stability of systems [23]. Macro-encapsulated PCM can be designed with different geometries mainly based on rectangular [24], cylindrical [25,26], and spherical shapes [27] that can be adapted to different applications. The effect of the design of macro-encapsulation on the heat transfer performance is mostly analyzed by numerical analysis with only a few experimental studies available in the literature, highlighting a research gap. Amongst the experimental studies available, Erlbeck et al. [28] and Al-Yasiri and Szabó [29] experimentally investigated the thermal behavior of concrete blocks with different shapes of microencapsulated PCM. Ismair and Moraes [30] numerically and experimentally evaluated spherical containers made with different geometries and materials and filled with PCM for cold storage domestic applications. This paper experimentally analyzes the effect of two different geometries of macro-encapsulated PCM in rectangular slabs on the performance of an energy storage tank. The analyzed TES tank is part of the generic heating system designed for the EU funded project SWS-HEATING (GA 764025). In particular, the PCM tank is used in the system as a thermal buffer to store the solar energy at low-grade temperature (15 ± 5 °C) to be supplied to a novel seasonal TES based on selective water sorbent materials. To the best of the authors knowledge, very few experimental studies on PCM tanks with rectangular slabs were published in the literature. One of the first papers was published by Moreno et al. [31] in which the performance of a TES tank filled with commercial PCM encapsulated in rectangular slabs was compared with the same tank filled with water. The results showed that the energy storage capacity of the tank filled with PCM was increased by 35.5% compared with same tank filled with water. Another study published by D'Avignon and Kummert [32] reported the results of experimental tests performed to study the behavior of a real-scale PCM storage at different operating conditions. One of the main conclusions from the study was that the PCM hysteresis and sub-cooling effects deviate the expected behavior from the experimental results. Liu et al. [33] used the experimental results obtained from the testing of a PCM tank filled with rectangular slabs containing a PCM with a sub-zero melting temperature (−26.7 °C) suitable for refrigerated transport, and glycol as heat transfer fluid. The developed model was based on a one-dimensional approach considering the temperature variations along direction of the heat transfer fluid showing a good agreement with the test. All experimental studies mentioned were carried out using a fixed design of the PCM tank without changing any boundaries related to the geometry or the configuration of the storage tank.

However, the geometrical design of the PCM encapsulation has a large influence on the thermal behavior of the PCM affecting the melting and the solidification process, and consequently the heat transfer [34]. In the case of rectangular shapes, the aspect ratio (height to width ratio) is a parameter that has to be taken into account in the design of TES tanks [21]. This paper shows for the first time a comparison based on experimental results of the thermal behavior of two different designs of macro-encapsulation of rectangular PCM slabs. The behavior of a thermal energy storage tank was analyzed using commercial PCM slabs with different thicknesses. The comparison of the two designs was done in terms of temperature profile, heat transfer rate, and energy obtained during the discharging process.

The main results obtained from the experimental tests reported in this paper can be used as a reference for institutions and manufacturers to optimize future designs of PCM tanks.

2. Materials and Methods

2.1. Materials

The PCM selected in this experimentation was PlusICE S15 (hydrated salt), supplied by PCM products, United Kingdom [35]. The main thermophysical properties of this material are shown in Table 1. Moreover, water was used as the heat transfer fluid (HTF).

Table 1. Thermophysical properties of PlusICE S15 [35].

Properties	Value
Melting temperature [°C]	15
Latent heat [J/g]	180
Specific heat capacity [kJ/kg·K]	1.90
Density [kg/m^3]	1700–1800
Thermal conductivity [W/(m·K)]	0.43
Maximum operation temperature [°C]	60

2.2. Experimental Set-Up

The experiments presented in this paper were carried out at the laboratory of the GREiA research group at the University of Lleida in Spain, in a set-up designed to test and characterize latent heat TES systems for mid-low temperature applications (-20 °C < T < 100 °C). Figure 1 shows a detailed schematic diagram of the experimental set-up composed by a 25 L inertia water tank, whose temperature is controlled by a vapor compression cooling unit (Zanotti model GCU2030ED01B [36]) of 5 kW cooling power, two immersion thermostats (OVAN TH100E-2kW [37], and JP SELECTA-1kW [38]). The set-up also integrates: two variable speed pumps, used to control the flow and inlet temperature at the TES system; and a flow meter Badger meter type ModMAG M1000 [39] with an accuracy of $\pm$0.25 % of the actual value, and the latent heat TES storage. The connections between components were joined using 0.5" diameter copper pipes insulated with 18 $\times$ 0.9 mm polyurethane tubes. The data acquisition system used consisted of 3 STEP DL-01 data logger [40] connected to a computer that integrates a system control and data acquisition software (SCADA) developed in InduSoft Web Studio [41]. The data recording interval was set to 10 s.

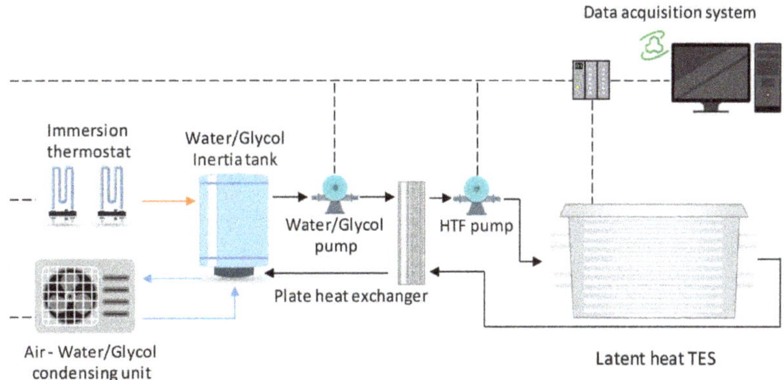

Figure 1. Schematic view of the experimental set-up used to perform the experimentation.

Figure 2 shows the PCM storage tank connected to the experimental set-up. The tests were carried out with two different PCM macro-encapsulation designs, namely, ThinICE and FlatICE (Figure 3). The containers were made in HDPE. The external dimensions of each design are reported in Figure 3, characterized by presenting similar length and width (A and B), but different thickness, with FlatICE dimensions being double that of ThinICE (C). Furthermore, the (D) dimension reveals that the use of thin macro-encapsulation enabled a larger distance between the slabs, increasing the space that allows circulating the HTF through the TES tank. Considering the aforementioned dimensions shown in Figure 3, the use of ThinICE encapsulation allowed fitting a larger number of slabs inside the tank, but less amount of latent storage material compared with the FlatICE, as shown in Table 2.

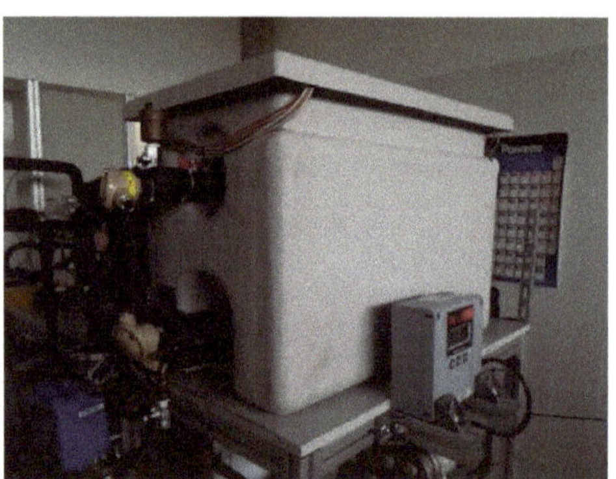

Figure 2. Latent heat TES.

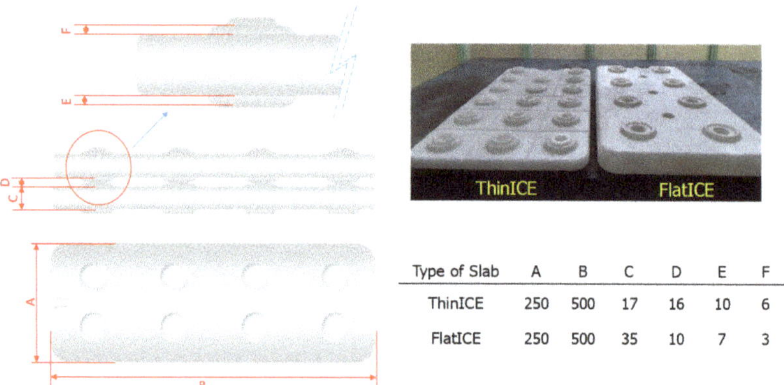

Type of Slab	A	B	C	D	E	F
ThinICE	250	500	17	16	10	6
FlatICE	250	500	35	10	7	3

Figure 3. ThinICE and FlatICE slabs encapsulation. Dimensions in millimeters.

Table 2. Total weight of PCM inside the tank.

Properties	FlatICE	ThinICE
Capacity of the slab [liters]	3.2	1.7
Weight of the container [kg]	0.55	0.5
Total weight of a single slab [kg]	6.7	3.8
Number of slabs inside the PCM tank	10	13
Total amount of PCM inside the tank [kg]	61.5	41.9
Total weight of the slabs inside the tank [kg]	67	49.4

The temperature inside the PCM storage was measured using nine Pt-100 class B, IEC 60751 standard type, with an accuracy of (0.3 + 0.005 · T). The sensors were fixed as shown in Figure 4 to the external surface of three different PCM slabs placed at the bottom, middle, and top of the tank, respectively. Moreover, two additional Pt-100 class A IEC 60751 standard type with an accuracy of (0.15 + 0.002 · T) sensors were placed at the inlets and outlets of the storage tank.

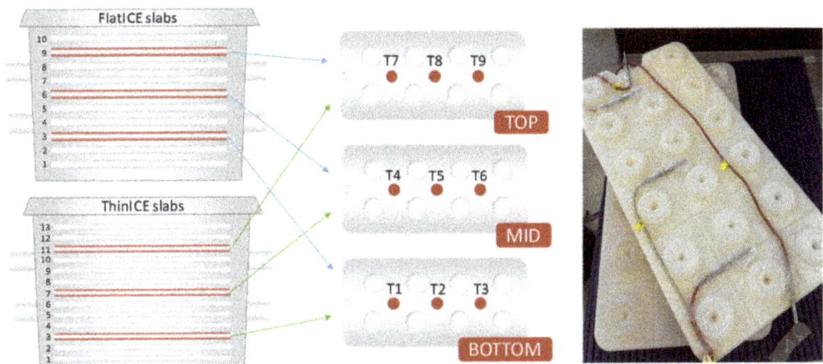

Figure 4. Temperature sensors location inside the storage tank.

2.3. Methodology

The experimental tests consisted of performing four different charging and discharging processes to evaluate the effect of the PCM macro-encapsulation design and the flow rate on the temperature distribution, heat transfer rate, and energy stored/released. At least three repetitions of each process were performed to ensure repeatability. A summary of the flow rates and temperatures used in the experimentation is shown in Table 3. Furthermore, the heat losses in the worst-case scenario analyzed represent 4% of the charging/discharging energy, therefore the analysis of heat losses was not included in the paper.

Table 3. Summary of the main parameters of the processes.

Process	Slab Type	Flow Rate [L/min]	HTF Inlet Temperature [°C]	PCM Tank Average Initial Temperature [°C]	Code
Charge	ThinICE	2	25	5 ± 1	C_ThinICE_2L
Charge	ThinICE	4	25	5 ± 1	C_ThinICE_4L
Charge	FlatICE	2	25	5 ± 1	C_FlatICE_2L
Charge	FlatICE	4	25	5 ± 1	C_FlatICE_4L
Discharge	ThinICE	2	5	25 ± 1	D_ThinICE_2L
Discharge	ThinICE	4	5	25 ± 1	D_ThinICE_4L
Discharge	FlatICE	2	5	25 ± 1	D_FlatICE_2L
Discharge	FlatICE	4	5	25 ± 1	D_FlatICE_4L

To perform a charging process, HTF was first circulated through the PCM tank until all sensors inside the tank reached a temperature of 5 ± 1 °C. Then, the HTF inlet temperature was set at 25 ± 1 °C and the flow rate was set to the corresponding value of the experiment shown in Table 3. The charging process was considered complete when the HTF temperature at the outlet of the tank reached 25 °C. To perform a discharging process, HTF was first circulated through the PCM tank until all sensors inside the tank reached a temperature of 25 ± 1 °C. Then, the HTF inlet temperature was set at 5 ± 1 °C and the flow rate was set to the corresponding value of the experiment in Table 3. The discharging process was considered complete when the HTF at the outlet of the tank reached 7 °C. This value was used instead of 5 °C because a minimum temperature difference of 2 °C was assumed between inlet and outlet of the storage tank as a constraint from the demand side.

2.4. Uncertainties Analysis

The impact of the uncertainties in the calculated parameters from the different measurements was evaluated by performing an uncertainty analysis using the Kline McClintock method. The uncertainties of the different monitored parameters are shown in Table 4. HTF specific heat capacity and density were calculated following the correlations presented in Equations (1) and (2) [42]:

$$\rho_{HTF} = 1.38 \cdot 10^{-5} \cdot T_{HTF}^3 - 5.63 \cdot 10^{-3} \cdot T_{HTF}^2 + 3.6 \cdot 10^{-3} \cdot T_{HTF}^1 + 1000 \tag{1}$$

$$Cp_{HTF} = 2.69 \cdot 10^{-9} \cdot T_{HTF}^4 - 6.63 \cdot 10^{-7} \cdot T_{HTF}^3 + 6.67 \cdot 10^{-5} \cdot T_{HTF}^2 - 2.67 \cdot 10^{-3} \cdot T_{HTF}^1 + 4.21 \tag{2}$$

Table 4. Uncertainties of the different parameters involved in the analyses of the present study.

Parameter	Units	Sensor	Accuracy
Temperature	°C	Pt-100 1/5 DIN class B IEC 60751	$\pm 0.3 + 0.005 \cdot T$
Temperature	°C	Pt-100 1/5 DIN class A IEC 60751	$\pm 0.15 + 0.002 \cdot T$
Flow rate	L/min	Badger meter type ModMAG M1000	$\pm 0.25\%$

By applying Equation (3) to the different parameters [43], the uncertainties of the HTF thermophysical properties (density and specific heat) as well as of the heat transfer rates and total stored/released energy were estimated. The uncertainty of the HTF thermophysical properties and heat transfer rates was estimated at each registered time step, and then the mean value was used. Table 5 shows the average uncertainties of the HTF density, specific heat, heat transfer rate, and stored/released energy during the different processes carried out:

$$W_R = \left[\left(\frac{\partial R}{\partial x_1} \cdot w_{x_1} \right)^2 + \left(\frac{\partial R}{\partial x_2} \cdot w_{x_2} \right)^2 + \cdots + \left(\frac{\partial R}{\partial x_n} \cdot w_{x_n} \right)^2 \right]^{1/2} \tag{3}$$

where W_R is the estimated uncertainty in the final result, R the function which depends on the measured parameters, x_n is the different independent monitored parameters, and w_x is the uncertainties associated to those independent parameters.

Table 5. Estimated uncertainties of the HTF thermophysical properties, heat transfer rate, and cumulated energy.

Test	Density [$\pm$kg/m^3]	Specific Heat [$\pm$kJ/kg · °C]	Heat Transfer Rate [$\pm$kW]	Accumulated Energy [$\pm$kJ]
C_ThinICE_2L	± 1.28	$\pm 2.44 \cdot 10^{-2}$	± 0.039	± 19
C_ThinICE_4L	± 1.28	$\pm 2.42 \cdot 10^{-2}$	± 0.055	± 19
C_FlatICE_2L	± 1.27	$\pm 2.43 \cdot 10^{-2}$	± 0.038	± 23
C_FlatICE_4L	± 1.28	$\pm 2.45 \cdot 10^{-2}$	± 0.054	± 24
D_ThinICE_2L	$\pm 5.5 \cdot 10^{-2}$	$\pm 2.95 \cdot 10^{-3}$	± 0.039	± 19
D_ThinICE_4L	$\pm 5.6 \cdot 10^{-2}$	$\pm 3.05 \cdot 10^{-3}$	± 0.055	± 20
D_FlatICE_2L	$\pm 5.3 \cdot 10^{-2}$	$\pm 2.77 \cdot 10^{-3}$	± 0.039	± 23
D_FlatICE_4L	$\pm 5.6 \cdot 10^{-2}$	$\pm 3.21 \cdot 10^{-3}$	± 0.055	± 23

3. Results and Discussion

3.1. Temperature Evolution during the Charging Process

Figure 5 shows the charging temperature profile of all sensors placed at the surface of the slabs for the two PCM encapsulation design at different flow rates. To analyze the effect of the encapsulation design in the charging duration, both slabs types were compared at the same flow rate. Due to the higher heat transfer surface and the reduced amount of PCM (30% less according to Table 3) when using ThinICE slabs, at both flow rates, the experiment with the ThinICE slabs reached full charge (T_out = 25 °C) 14% faster than with FlatICE. Furthermore, when analyzing the impact of the flow rate, in both slab designs the experiments show that at 4 L/min the full charge is reached 60% faster than at 2 L/min.

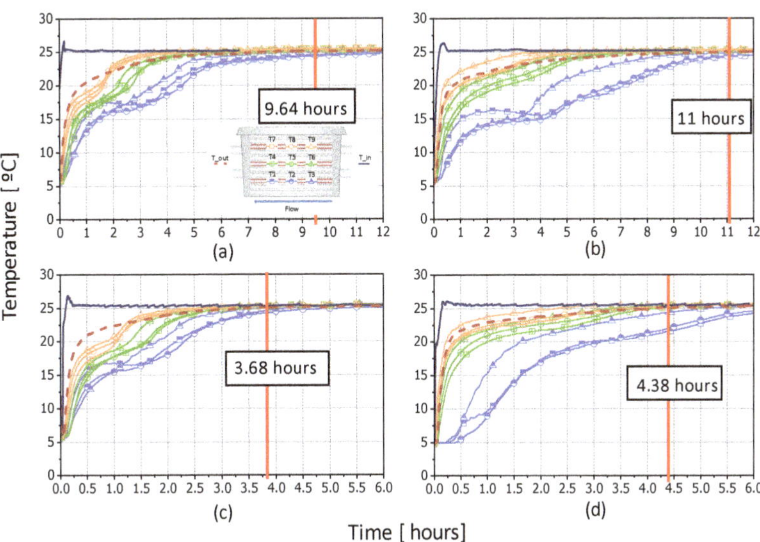

Figure 5. Charge PCM slab temperature profile for different slabs and HTF flow rates: (**a**) C_ThinICE and 2 L/min, (**b**) C_FlatICE and 2 L/min, (**c**) C_ThinICE and 4 L/min, and (**d**) C_FlatICE and 4 L/min. Note: The red line denotes the end of the charging experiment (T_out = 25 °C). The time axis is not presented on the same scale in all the figures.

When comparing temperature distribution inside the tank (Figure 5), a constant stratification profile between the top, middle, and bottom slabs was observed in all the experiments. This effect was more pronounced in the tests performed with FlatICE slabs in which the lower part of the tank takes longer to charge, obtaining a temperature gradient up to 15 K between the coldest and hottest regions of the tank. This can be explained by the fact that the tank with FlatICE fits a lower number of slabs, as well as presenting smaller HTF channels compared with the tank with the ThinICE design (Figure 3, Table 2). Therefore, in this tank, the opposition to the HTF flow is higher, enhancing the distribution of the latter towards the regions of the tank where the density is more similar to the HTF inlet one (i.e., upper and middle region of the tank).

3.2. Heat Transfer Rate Evolution and Total Energy Stored in the Charging Process

Figure 6 presents the evolution of the heat transfer rate (HTR) during the charging process of the four studied cases. Due to the characteristics of the experimental set-up, at the beginning of the experiment, the inlet temperature of the tank oscillated ±2 °C with respect to the desired temperature, affecting the initial peak of the heat transfer rate. However, the inlet temperature stabilized (with Tin standard deviation lower than 0.3) before the temperature inside the tank reaches the latent range of the PCM. The HTR profiles showed an exponential behavior with significantly higher values during the first 20 min of the process when the heat is mainly transferred to the HTF inside the tank and, therefore, rapidly increases its temperature. Afterwards, while the PCM temperature increases, the values of the heat transfer exponentially decrease until minimum values.

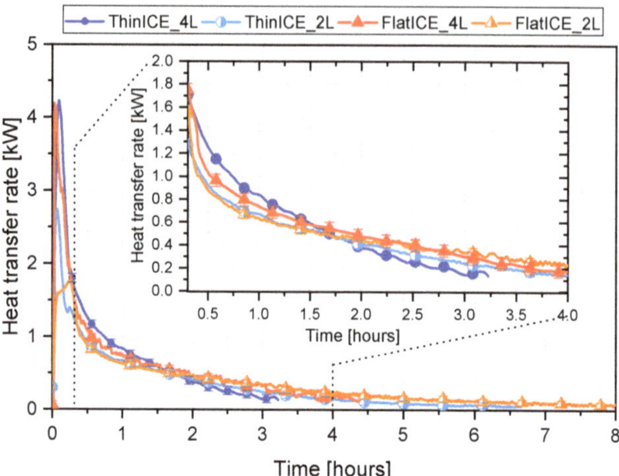

Figure 6. Evolution of the HTF heat transfer rate during the charging processes of the four study cases presented in this study.

During the first 1.5 h of operation, ThinICE_2L and FlatICE_2L showed similar HTR values, which indicates that, due to the low flow rate, the heat transfer by convection is low. Therefore, the higher heat transfer surface area existing with ThinICE slabs is not fully exploited. Moreover, after 1.5 h the HTR delivered to the ThinICE_2L decreases faster than the one delivered to the FlatICE_2L due to the higher amount of PCM introduced into the tank with FlatICE slabs. At a higher flow rate, heat transfer by convection increases. Therefore, during the first 1.5 h of operation, the bigger heat transfer surface area present with ThinICE_4L slabs increases its HTR over FlatICE_4L. After this period, and similar to the results at 2 L/min, the power delivered to the ThinICE_4L decreased faster than the one delivered to the FlatICE_4L.

When analyzing the effect of the flows in each slab type, the influence is greater in the tank with ThinICE slabs obtaining, after the initial peak, up to 0.4 kW more in ThinICE_4L than in ThinICE_2L. In the case of the tank with FlatICE, this increase drops to 0.1 kW when comparing FlatICE_4L vs FlatICE_2L. The latter results corroborate the statement above; the increase in heat transfer by convection, as the flow rate increases, is more pronounced in the tank with ThinICE slabs due to the larger heat transfer surface and the lower thickness of the PCM layer using this type of slab.

Figure 7 reports the total energy stored for each experiment condition. The results with ThinICE slabs show that the flow variation did not affect the total stored energy, suggesting the correct utilization of the energy storage capacity of the PCM. Conversely, when analyzing the tank with FlatICE slabs, the charging experiments at 4 L/min stored 10% less energy than the same experiment at 2 L/min. This is due to changes in the flow rate distribution between the slab channels inside the tank when increasing the flow rate. At the end of the experiment, (T_out 25 °C) with FlatICE at 4 L/min, the PCM in the bottom slabs of the tank had not completed the phase change (Figure 5) and therefore stored 8% less energy.

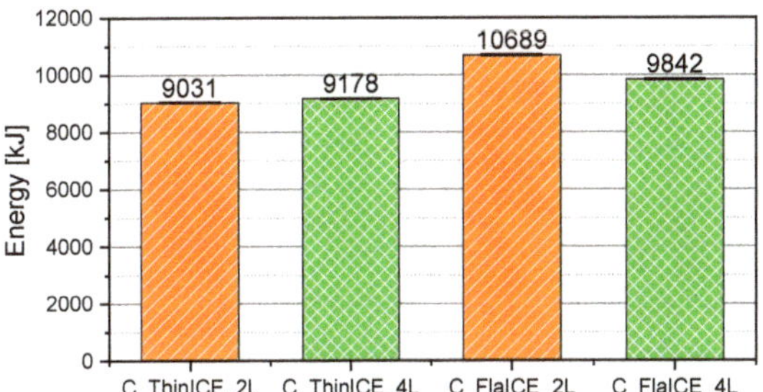

Figure 7. Total energy delivered to the PCM storage tank during the charging processes of the four study cases presented in this study.

3.3. Temperature Evolution during the Discharging Process

Figure 8 shows the discharging temperature profile of all sensors placed at the surface of the slabs at two different mass flow rates. Analyzing the influence of the PCM encapsulation design on both flow rates, the tank with ThinICE slabs finished the discharging process 30% faster than with FlatICE slabs. This can be explained by the higher heat transfer surface and the lower amount of PCM (30% less according to Table 3) when using ThinICE. Moreover, when analyzing the influence of the flow rate, Figure 8 shows that for both slab types at 4 L/min the experiments were completed 50% faster than at 2 L/min.

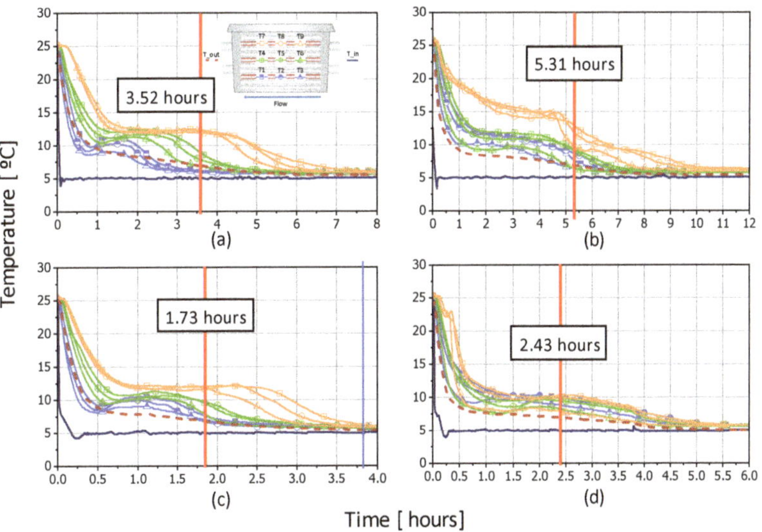

Figure 8. Discharge process PCM slab temperature profile for different slabs and HTF flow rates: (**a**) ThinICE and 2 L/min, (**b**) FlatICE and 2 L/min, (**c**) ThinICE and 4 L/min, and (**d**) FlatICE and 4 L/min. Note: The red line denotes the end of the charging experiment (T_out = 25 °C). The time axis is not presented on the same scale in all the figures.

When comparing temperature distribution inside the tank (Figure 8), a similar temperature profile between the top, middle, and bottom slabs was observed in all the experiments performed with ThinICE slabs. Moreover, this behavior changed with the use of the FlatICE, where the stratification and the profile temperature inside the tank depends on the mass flow rate.

3.4. Heat Transfer Rate Evolution and Total Energy Released in the Discharging Process

The HTR evolution during the discharging process for all the experimental cases is shown in Figure 9. In all the experiments the profiles showed a similar trend. Significantly higher values were obtained during the first 20 min of the process when the heat is mainly transferred from the HTF inside the tank followed by an exponential decrease while the PCM decreases its temperature until minimum values are reached. Furthermore, in this case, due to the characteristics of the experimental facility, at the beginning of the experimentation the inlet temperature of the tank oscillates ±2 °C around the desired temperature, affecting the initial peak of power. However, the inlet temperature stabilizes (T_in standard deviation lower than 0.3 °C) before the temperature inside the tank reaches the latent range of the PCM. After the initial peak and during the first 1.5 h of operation, ThinICE_2L shows slightly better performance getting up to 0.1 kW more HTR than FlatICE_2L. Moreover, due to the lower amount of PCM in the storage tank with ThinICE slabs, after 1.5 h the HTR delivered by ThinICE_2L decreases faster than the one delivered by FlatICE_2L. At 4 L/min, after the initial peak and during the first 1.5 h, similar results to 2 L/min are obtained. Moreover, after 1.5 h the HTR of ThinICE_4L drastically decreases, therefore for the next 2 h (from 1.5–3.5 h) FlatICE_4L maintains an HTR up to 0.4 kW higher than ThinICE_2L.

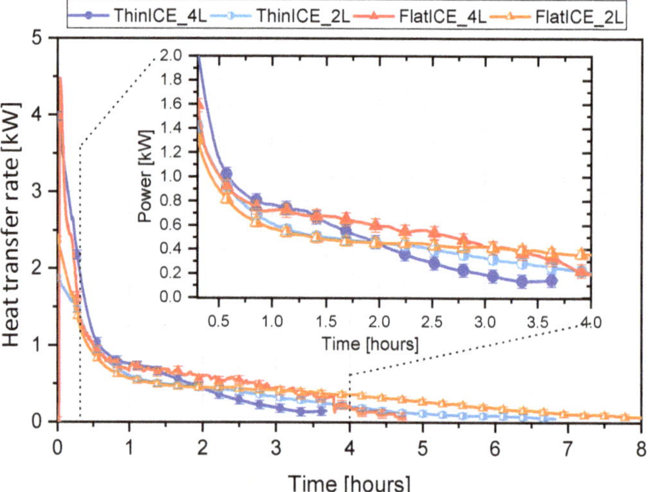

Figure 9. Evolution of the HTF heat transfer rate during the discharging processes of the four study cases presented in this study.

Figure 10 reports the total energy released for each experiment conditions, and the percentage it represents with respect to the energy stored in the charging process (Figure 7) in each case. Analyzing the effects of flow rate, similar to the charging, both experiments with ThinICE slabs showed a comparable energy release, suggesting a correct utilization of the energy stored in the PCM. In the case of FlatICE slabs, experiments at 2 L/min released 10% more energy compared with 4 L/min. This is supported by the fact that in the charging process the tank at 2 L/min manages to store 10% more energy than at 4 L/min (Figure 5). In addition, it is interesting to note that in all cases of the selected operating threshold approximately 85% of the energy stored in the tank was discharged.

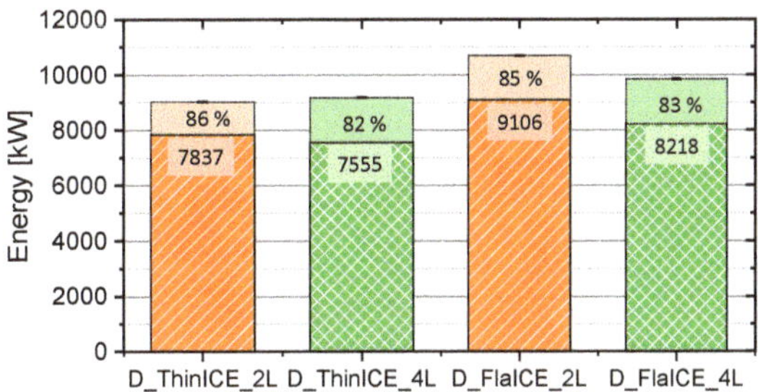

Figure 10. Total energy released by the PCM storage tank during the discharging processes of the four study cases presented in this study.

4. Conclusions

Macro-encapsulation of phase change materials (PCM) represents one of the most widely used techniques for the implementation of latent heat thermal energy storage systems. The design of the macro-encapsulation is fundamental to archive the best compromise between optimal heat transfer performance and energy stored. However, current literature lacks experimental data on the effect of macro-encapsulation in the performance of latent heat thermal energy storage.

This paper analyzed, through an experimental study, the effect of the design of macro-encapsulated PCM on the thermal behavior of a latent heat thermal energy storage tank during both the charging and discharging processes. In this study, external dimensions of the energy storage tank were fixed and two different types of commercial slabs with different thickness filled with the same PCM were tested. The results could be particularly useful to evaluate the best configuration of storage medium when the storage tank is limited with a fixed volume.

The results were compared in terms of temperature profile, heat transfer rate, and energy stored/released. The results and the conclusions obtained from this study can be applied to similar configuration of the PCM storage that aim to use rectangular macro-encapsulated slabs as storage medium. The lesson learnt from this study suggests that macro-encapsulation design has a relevant impact on the heat transfer during both charging and discharging processes, so the design of the TES unit should be done and analyzed according to the requirements of the application.

The use of a thinner macro-encapsulation design (ThinICE) allowed fitting a larger number of slabs inside the tank. However, the higher amount of encapsulation material and the larger distance between the slabs (i.e., higher HTF channels height) resulted in a 30% less amount of PCM introduced inside the tank with this encapsulation design.

With ThinICE slabs, the temperature profiles were less affected by the influence of the mass flow rate, promoting a stratified temperature profile inside the tank in both the charging and discharging processes. Using FlatICE, this effect is more pronounced at low flow rates due to the smaller height of the channels that obstructed the flow at the bottom of the tank during charging and at the top of the tank during discharging. However, at high flow rates, the stratification is reduced with the use of thicker slabs, especially during the discharging process.

In all the discharging tests, when the outlet temperature of the tank reached 7 °C, approximately 85% of the energy previously stored in the tank was discharged.

The effect of increasing the heat transfer surface using ThinICE slabs on the power delivered by the storage tank is mostly appreciated at a higher flow rate where the heat transferred by convection is higher. Furthermore, using thinner slabs, the higher heat transfer surface area achieves a higher discharging power but is delivered for a shorter period of time. Therefore, for longer discharging periods and for higher storage capacity given a fixed volume of storage tank, the use of FlatICE should be preferred.

Author Contributions: Conceptualization, D.V., E.B., and L.F.C.; methodology, D.V., E.B., and G.Z.; formal analysis, D.V., E.B., A.C., B.D.M., Á.d.G., G.Z., and L.F.C.; investigation, D.V., A.C., and E.B.; resources, L.F.C.; data curation, L.F.C.; writing—original draft preparation, D.V. and E.B.; writing—review and editing, D.V., E.B., A.C., B.D.M., Á.d.G., and G.Z., L.F.C.; visualization, D.V.; supervision, L.F.C.; project administration, L.F.C.; and funding acquisition, L.F.C. All authors have read and agreed to the published version of the manuscript.

Funding: This project has received funding from the European Union's Horizon 2020 research and innovation program under grant agreement No. 764025 (SWS-HEATING). This work was partially funded by the Ministerio de Ciencia, Innovación y Universidades de España (RTI2018-093849-B-C31—MCIU/AEI/FEDER, UE) and by the Ministerio de Ciencia, Innovación y Universidades—Agencia Estatal de Investigación (AEI) (RED2018-102431-T). This work is partially supported by ICREA under the ICREA Academia program.

Institutional Review Board Statement: Not applicable.

Informed Consent Statement: Not applicable.

Data Availability Statement: The data presented in this study are available on request from the corresponding author.

Acknowledgments: The authors would like to thank the Catalan Government for the quality accreditation given to their research group (2017 SGR 1537). GREiA is a certified agent TECNIO in the category of technology developers from the Government of Catalonia. Boniface Dominick Mselle would like to thank Programa Santander PredocUdL for his research fellowship. Alicia Crespo would like to acknowledge the financial support of the FI-SDUR grant from the AGAUR of the Generalitat de Catalunya and Secretaria d'Universitats i Recerca del Departament d'Empresa i Coneixement de la Generalitat de Catalunya.

Conflicts of Interest: The authors declare no conflict of interest.

References

1. Borri, E.; Zsembinszki, G.; Cabeza, L.F. Recent developments of thermal energy storage applications in the built environment: A bibliometric analysis and systematic review. *Appl. Therm. Eng.* **2021**, *189*, 116666. [CrossRef]
2. Navarro, L.; de Gracia, A.; Colclough, S.; Browne, M.; McCormack, S.J.; Griffiths, P.; Cabeza, L.F. Thermal energy storage in building integrated thermal systems: A review. Part 1. active storage systems. *Renew. Energy* **2016**, *88*, 526–547. [CrossRef]
3. Navarro, L.; de Gracia, A.; Niall, D.; Castell, A.; Browne, M.; McCormack, S.J.; Griffiths, P.; Cabeza, L.F. Thermal energy storage in building integrated thermal systems: A review. Part 2. Integration as passive system. *Renew. Energy* **2016**, *85*, 1334–1356. [CrossRef]
4. Heier, J.; Bales, C.; Martin, V. Combining thermal energy storage with buildings—A review. *Renew. Sustain. Energy Rev.* **2015**, *42*, 1305–1325. [CrossRef]
5. Mselle, B.D.; Vérez, D.; Zsembinszki, G.; Borri, E.; Cabeza, L.F. Performance study of direct integration of phase change material into an innovative evaporator of a simple vapour compression system. *Appl. Sci.* **2020**, *10*, 4649. [CrossRef]
6. Palomba, V.; Bonanno, A.; Brunaccini, G.; Aloisio, D.; Sergi, F.; Dino, G.E.; Varvaggiannis, E.; Karellas, S.; Nitsch, B.; Strehlow, A.; et al. Hybrid cascade heat pump and thermal-electric energy storage system for residential buildings: Experimental testing and performance analysis. *Energies* **2021**, *14*, 2580. [CrossRef]
7. Romaní, J.; Belusko, M.; Alemu, A.; Cabeza, L.F.; de Gracia, A.; Bruno, F. Control concepts of a radiant wall working as thermal energy storage for peak load shifting of a heat pump coupled to a PV array. *Renew. Energy* **2018**, *118*, 489–501. [CrossRef]
8. Wang, J.; Zhang, Q.; Yu, Y.; Chen, X.; Yoon, S. Application of model-based control strategy to hybrid free cooling system with latent heat thermal energy storage for TBSs. *Energy Build.* **2018**, *167*, 89–105. [CrossRef]
9. Yu, C.-R.; Guo, H.-S.; Wang, Q.-C.; Chang, R.-D. Revealing the impacts of passive cooling techniques on building energy performance: A residential case in Hong Kong. *Appl. Sci.* **2020**, *10*, 4188. [CrossRef]
10. Kameni Nematchoua, M.; Vanona, J.C.; Orosa, J.A. Energy efficiency and thermal performance of office buildings integrated with passive strategies in coastal regions of humid and hot tropical climates in Madagascar. *Appl. Sci.* **2020**, *10*, 2438. [CrossRef]
11. Prieto, C.; Fereres, S.; Cabeza, L.F. The role of innovation in industry product deployment: Developing thermal energy storage for concentrated solar power. *Energies* **2020**, *13*, 2943. [CrossRef]
12. Palomba, V.; Brancato, V.; Frazzica, A. Experimental investigation of a latent heat storage for solar cooling applications. *Appl. Energy* **2017**, *199*, 347–358. [CrossRef]
13. Guelpa, E.; Verda, V. Thermal energy storage in district heating and cooling systems: A review. *Appl. Energy* **2019**, *252*, 113474. [CrossRef]
14. Miró, L.; Gasia, J.; Cabeza, L.F. Thermal energy storage (TES) for industrial waste heat (IWH) recovery: A review. *Appl. Energy* **2016**, *179*, 284–301. [CrossRef]

15. Crespo, A.; Barreneche, C.; Ibarra, M.; Platzer, W. Latent thermal energy storage for solar process heat applications at medium-high temperatures—A review. *Sol. Energy* **2019**, *192*, 3–34. [CrossRef]
16. Borri, E.; Sze, J.Y.; Tafone, A.; Romagnoli, A.; Li, Y.; Comodi, G. Experimental and numerical characterization of sub-zero phase change materials for cold thermal energy storage. *Appl. Energy* **2020**, *275*, 115131. [CrossRef]
17. Mehling, H.; Cabeza, L.F. *Heat and Cold Storage with PCM. An Up to Date Introduction into Basics and Applications*, 1st ed.; Springer: Berlin/Heidelberg, Germany, 2008; ISBN 979-3-540-68556-2.
18. Gasia, J.; Tay, N.H.S.; Belusko, M.; Cabeza, L.F.; Bruno, F. Experimental investigation of the effect of dynamic melting in a cylindrical shell-and-tube heat exchanger using water as PCM. *Appl. Energy* **2017**, *185*, 136–145. [CrossRef]
19. Huang, K.; Liang, D.; Feng, G.; Jiang, M.; Zhu, Y.; Liu, X.; Jiang, B. Macro-encapsulated PCM cylinder module based on paraffin and float stones. *Materials* **2016**, *9*, 361. [CrossRef]
20. Besagni, G.; Croci, L. Experimental study of a pilot-scale fin-and-tube phase change material storage. *Appl. Therm. Eng.* **2019**, *160*, 114089. [CrossRef]
21. Liu, Z.; Yu, Z.J.; Yang, T.; Qin, D.; Li, S.; Zhang, G.; Haghighat, F.; Joybari, M.M. A review on macro-encapsulated phase change material for building envelope applications. *Build. Environ.* **2018**, *144*, 281–294. [CrossRef]
22. Khudhair, A.M.; Farid, M.M. A review on energy conservation in building applications with thermal storage by latent heat using phase change materials. *Energy Convers. Manag.* **2004**, *45*, 263–275. [CrossRef]
23. Chandel, S.S.; Agarwal, T. Review of current state of research on energy storage, toxicity, health hazards and commercialization of phase changing materials. *Renew. Sustain. Energy Rev.* **2017**, *67*, 581–596. [CrossRef]
24. Sun, X.; Chu, Y.; Medina, M.A.; Mo, Y.; Fan, S.; Liao, S. Experimental investigations on the thermal behavior of phase change material (PCM) in ventilated slabs. *Appl. Therm. Eng.* **2019**, *148*, 1359–1369. [CrossRef]
25. Carmona, M.; Rincón, A.; Gulfo, L. Energy and exergy model with parametric study of a hot water storage tank with PCM for domestic applications and experimental validation for multiple operational scenarios. *Energy Convers. Manag.* **2020**, *222*, 113189. [CrossRef]
26. Xu, T.; Humire, E.N.; Chiu, J.N.-W.; Sawalha, S. Numerical thermal performance investigation of a latent heat storage prototype toward effective use in residential heating systems. *Appl. Energy* **2020**, *278*, 115631. [CrossRef]
27. Goeke, J.; Schwamborn, E. Phase change material in spherical capsules for hybrid thermal storage. *Chem. Ing. Tech.* **2020**, *92*, 1098–1108. [CrossRef]
28. Erlbeck, L.; Schreiner, P.; Schlachter, K.; Dörnhofer, P.; Fasel, F.; Methner, F.J.; Rädle, M.A. of thermal behavior by changing the shape of P. inclusions in concrete blocks Adjustment of thermal behavior by changing the shape of PCM inclusions in concrete blocks. *Energy Convers. Manag.* **2018**, *158*, 256–265. [CrossRef]
29. Al-Yasiri, Q.; Szabó, M. Thermal performance of concrete bricks based phase change material encapsulated by various aluminium containers: An experimental study under Iraqi hot climate conditions. *J. Energy Storage* **2021**, *40*, 102710. [CrossRef]
30. Ismail, K.A.R.; Moraes, R.I.R. A numerical and experimental investigation of different containers and PCM options for cold storage modular units for domestic applications. *Int. J. Heat Mass Transf.* **2009**, *52*, 4195–4202. [CrossRef]
31. Moreno, P.; Castell, A.; Solé, C.; Zsembinszki, G.; Cabeza, L.F. PCM thermal energy storage tanks in heat pump system for space cooling. *Energy Build.* **2014**, *82*, 399–405. [CrossRef]
32. D'Avignon, K.; Kummert, M. Experimental assessment of a phase change material storage tank. *Appl. Therm. Eng.* **2016**, *99*, 880–891. [CrossRef]
33. Liu, M.; Saman, W.; Bruno, F. Validation of a mathematical model for encapsulated phase change material flat slabs for cooling applications. *Appl. Therm. Eng.* **2011**, *31*, 2340–2347. [CrossRef]
34. Hosseini, M.J.; Ranjbar, A.A.; Sedighi, K.; Rahimi, M. A combined experimental and computational study on the melting behavior of a medium temperature phase change storage material inside shell and tube heat exchanger. *Int. Commun. Heat Mass Transf.* **2012**, *39*, 1416–1424. [CrossRef]
35. PCM Products. Available online: http://www.pcmproducts.net/ (accessed on 15 January 2019).
36. Zanotti. Available online: https://zanottiappliance.com/es/refrigeracion/refrigeracion-fija/ (accessed on 18 June 2021).
37. OVAN. Available online: https://ovan.es/ (accessed on 18 June 2021).
38. SELECTA. Available online: https://grupo-selecta.com/distribuidores/esli-sarl-engineering-scientific-lab-instruments/ (accessed on 18 June 2021).
39. Badgermeter. Available online: https://www.badgermeter.com/es-es/productos/medidores/medidores-de-flujo-electromagneticos/ medidor-de-flujo-electromagnetico-modmag-m1000/ (accessed on 18 June 2021).
40. Step Dl-01. Available online: https://store.stepsl.com/product?prod=DL-01 (accessed on 18 June 2021).
41. Indusoft Web Studio. Available online: https://www.aveva.com/en/products/indusoft-web-studio/ (accessed on 18 June 2021).
42. Kukulka, D. *Thermodynamic and Transport Properties of Pure and Saline Water*; State University of New York: Buffalo, NY, USA, 1981.
43. Holman, J. *Experimental Methods for Engineers*, 8th ed.; McGraw-Hill: New York, NY, USA, 2012; ISBN 0073529303.

Article

Thermal Energy Storage Performance of Tetrabutylammonium Acrylate Hydrate as Phase Change Materials

Hitoshi Kiyokawa [1], Hiroki Tokutomi [1], Shinichi Ishida [2], Hiroaki Nishi [3] and Ryo Ohmura [1,*]

[1] Department of Mechanical Engineering, Keio University, 3-14-1 Hiyoshi Kohoku-ku, Yokohama 223-8522, Japan; apalcoann@keio.jp (H.K.); konoMichini-0warihanaku@softbank.ne.jp (H.T.)

[2] Cyber Kobo LLC., 2-201, Kitabukuro-cho, Omiya-ku, Saitama-shi 330-0835, Japan; sin@cyber-lab.co.jp

[3] Department of System Design Engineering, Keio University, 3-14-1 Hiyoshi Kohoku-ku, Yokohama 223-8522, Japan; west@sd.keio.ac.jp

* Correspondence: rohmura@mech.keio.ac.jp; Tel.: +81-45-566-1813

Abstract: Kinetic characteristics of thermal energy storage (TES) using tetrabutylammonium acrylate (TBAAc) hydrate were experimentally evaluated for practical use as PCMs. Mechanical agitation or ultrasonic vibration was added to detach the hydrate adhesion on the heat exchanger, which could be a thermal resistance. The effect of the external forces also was evaluated by changing their rotation rate and frequency. When the agitation rate was 600 rpm, the system achieved TES density of 140 MJ/m^3 in 2.9 h. This value is comparable to the ideal performance of ice TES when its solid phase fraction is 45%. UA/V (U: thermal transfer coefficient, A: surface area of the heat exchange coil, V: volume of the TES medium) is known as an index of the ease of heat transfer in a heat exchanger. UA/V obtained in this study was comparable to that of other common heat exchangers, which means the equivalent performance would be available by setting the similar UA/V. In this study, we succeeded in obtaining practical data for heat storage by TBAAc hydrate. The data obtained in this study will be a great help for the practical application of hydrate heat storage in the future.

Keywords: lathrate hydrate; thermal energy storage; tetrabutylammonium acrylate (TBAAc); crystal growth; ultrasonic vibration

Citation: Kiyokawa, H.; Tokutomi, H.; Ishida, S.; Nishi, H.; Ohmura, R. Thermal Energy Storage Performance of Tetrabutylammonium Acrylate Hydrate as Phase Change Materials. *Appl. Sci.* **2021**, *11*, 4848. https://doi.org/10.3390/app11114848

Academic Editor: Ioannis Kartsonakis

Received: 17 April 2021
Accepted: 24 May 2021
Published: 25 May 2021

Publisher's Note: MDPI stays neutral with regard to jurisdictional claims in published maps and institutional affiliations.

1. Introduction

The use of sustainable, renewable energy sources is becoming increasingly important as global attention on environmental issues increases. However, the amount of electricity generated by renewable energy sources such as wind and solar power fluctuates greatly depending on the natural environment. In recent years, energy consumption has been increasing. It is essential to have a technology to fill the gap between the amount of electricity generated and the demand. These technologies would contribute to the load leveling [1].

In particular, the recent development in information and communication technology (ICT) (e.g., Internet of Things, 5G communications, cloud computing, big data) has produced a great demand for data centers (DCs). In 2020, the rapid spread of remote working due to the pandemic also boosted demand for DCs. The industrial use of these technologies is also driving this trend [2,3]. Masanet et al. [4] estimated that the electricity use in DCs accounted for 1% of worldwide electricity use in 2020. This trend will make DCs even more energy-intensive and will increase the cost of cooling the heat generated and the amount of electricity used [5]. Considering the generalization of ICT and the widespread use of the fifth-generation mobile communication system, the amount of communication will further increase worldwide. The importance of data centers will increase to support the exchange of huge amounts of data. A suitable cooling system will be able to keep the appropriate temperature. In addition, such systems are significant to realize the zero-downtime of a DC operation with low cost.

DC cooling systems are classified into two types. One is an air-based cooling system [6,7]. The other one is a liquid cooling system [8]. As for air-based system, the heat emitted from the computer is removed by air flow. The main drawback of this system is a low thermal conductivity, less than 0.03 W m^{-1} K^{-1} [9]. On the other hand, in a liquid cooling system, the heat is removed by the thermal conduction and convection into the liquid, mostly water. A liquid cooling system is inclined to be bigger and heavier because that system utilizes only sensible heat, which has low energy density. Thermal energy storage technology is used to store the heat energy with the heat capacity of substance. When the demand of the electricity is low, the surplus electricity is stored as cold energy. Then, the stored energy is changed to electricity when the demand increases, for example, daytime peak in power usage [10,11].

Phase Change Materials (PCMs) can be a competitive system for thermal energy storage because they can also utilize latent heat, which has a higher energy density [12–16]. Organic compounds and water are representative examples of PCMs for DC cooling. As for organic PCMs, they often have problems in terms of safety. For example, paraffin is known as an organic PCM, but it is not safe because of its flammability [17]. Water is not suitable because its phase change point is outside the range of operation temperature in DCs, 288 K to 305 K [18].

Utilizing clathrate hydrate for a thermal energy storage medium has been proposed as a better idea than water or an ice-based medium [19–23]. Clathrate hydrates, ice-like compounds, are formed when some gases have contact with water or ice under high pressure and/or low temperature [24]. The gases encapsulated in clathrate hydrates are called guest compounds. Clathrate hydrates have enormous potential for industrial usage thanks to their thermodynamic properties [19,22,25–28]. Considering the application for thermal energy storage medium, equilibrium temperature should be suitable for the DC operation temperature. Favorable equilibrium temperature can be obtained appropriately by selecting the guest compounds [29–31].

Until now, only tetrabutylammonium bromide (TBAB) hydrate has been commercially applied as a hydrate PCM for air conditioning. The thermodynamic properties of TBAB hydrate have been reported in the previous studies. TBAB hydrate has an equilibrium temperature at an atmospheric pressure of 285.9 K [32], a dissociation heat of 193 kJ/kg [33], and the thermal conductivity of 0.35 W m^{-1} K^{-1} [34]. As reported by ASHRAE Technical Committee [18], the appropriate temperature range of DC cooling is 288 K to 305 K. This is why TBAB hydrate is not favorable for DC cooling.

Sakamoto et al. [35] studied the thermodynamic properties of tetrabutylammonium acrylate (TBAAc) hydrate. According to the study, TBAAc hydrate had the highest equilibrium temperature of 291.5 K at $w_{\text{TBAAc}} = 0.36$, where w_{TBAAc} is the mass fraction of TBAAc. The greatest dissociation heat, a latent heat, of 195 kJ/kg was obtained at $w_{\text{TBAAc}} = 0.33$ under ambient pressure. As for the thermal conductivity of ionic semiclathrate hydrates, Fujiura et al. [34] studied the value of tetrabutylammonium bromide (TBAB) hydrate and tetrabutylammonium chloride (TBAC) hydrate, which have similar characteristics with TBAAc hydrate. They reported that TBAB hydrate and TBAC hydrate have thermal conductivity of approximately 0.40 W m^{-1} K^{-1}. According to the report by ASHRAE Technical Committee [18], the cooling system for DCs is required to be set from 288 K to 305 K. From the aspect of operation temperature in DCs, it can be said that TBAAc hydrates are suitable for utilizing in DC cooling as PCMs. Also, the dissociation heat of the TBAAc is larger than that of TBAB hydrate, which has been commercialized in the industrial area [33,35]. Given these advantages, TBAAc hydrates could be better PCMs for DC cooling. This is why it has significant meaning to practically evaluate the performance of TBAAc hydrates as PCMs. As a step for industrial use of TBAAc hydrate, we performed kinetic thermal energy storage experiments with TBAAc hydrate.

Based on these results, we obtained the data of energy storage density and energy storage rate during the experiments. During the experiments, we also found the adhesion of the hydrate to the surface of the heat exchanger. As the adhered hydrate is the thermal resistance to the further hydrate formation, we used an agitator or an ultrasonic transducer to break the adhesion. By comparing the data with and without external forces, such as the agitation and ultrasonic waves, we obtained the practical data for a hydrate-based thermal energy storage system in DC cooling.

2. Materials and Methods

2.1. Materials

The details on reagents used in this study were summarized in Table 1. Tetrabutylammonium Acrylate (TBAAc) aqueous solution was obtained by neutralizing Tetrabuthylammonium hydroxide (TBAOH) aqueous solution, 1.52 kg, (0.40 mass fraction, Sigma Aldrich Xo. LLC, Saint Louis, MO, USA) and Acrylic acid, 0.17 kg, (0.99 mass fraction in liquid reagent, Sigma Aldrich Xo. LLC, Saint Louis, State of MO, USA). Also, we added laboratory-made H_2O, 0.35 kg. The mass fraction of TBAAc (w_{TBAA_c})) was 0.36. The masses of all reagents were measured by an electronic balance (GF-600, A&D Co. Ltd., Tokyo, Japan) with an expanded uncertainty of $\pm$ 0.004 g (coverage factor, $k = 2$).

Table 1. Specification of the reagents used in this study.

Name	Chemical Formula	Supplier	Purity
Tetrabuthylammonium hydroxide	$(CH_3CH_2CH_2CH_2)_4NOH$	Sigma-Aldrich Co. LLC	0.40 mass fraction in aqueous solution
Acrylic acid	$CH_2=CHCOOH$	Sigma-Aldrich Co. LLC	0.99 mass fraction in liquid reagent
Tetrabuthylammonium Acrylate	$(CH_3CH_2CH_2CH_2)_4NOOCHC=CH_2$	Laboratory made from above solution	0.36 mass fraction in aqueous solution after the neutralization process. The standard uncertainty of mass fraction was $\pm 1.0\times10^{-4}$
Water	H_2O	Laboratory made	Electrical conductivity was less than 0.1 μS/cm

2.2. Apparatus

The schematic diagram of the experimental apparatus is illustrated in Figure 1. The insulating container's width, length and height are all 150 mm. Cooling water was circulated inside a copper coil tube (inner diameter 6.35 mm, outer diameter 4.35 mm) for heat exchange. This coil's diameter, pitch and number of turns are 100 mm, 10 mm, and 7 times, respectively. The temperature and flow rate in the coil was controlled by a chiller (CTP-3000 EYELA, Tokyo Rikakikai Co. Ltd., Tokyo, Japan) and flowmeter with a precision needle valve (RK 1250, KOFLOC Kyoto, Kyoto, Japan). As shown in Figure 1, the temperature of cooling water and TBAAc were measured by three platinum resistance temperature detectors (222-055, Electronic Temperature Instrument Ltd., West Sussex, UK) with an uncertainty of $\pm$ 0.1 K ($k = 2$). The surface of the coil would be covered with hydrate during the experiment. This hydrate prevents the heat exchange between the heat exchanger and thermal energy storage medium. We used an ultrasonic transducer (HEC-45282, HONDA Electronics Co. Ltd., Tokyo, Japan) or an agitator (SM-103D, Kenis Ltd., Osaka, Japan) to prevent the hydrate deposition.

2.3. Procedures

We made 2.0 kg of TBAAc aqueous solution by mixing the above reagents and put them in the container. The entire system was confirmed to be stationary at 298 K. T_{in} (outlet of the chiller) and flow rate, $\dot{V}_w$, of the cooling water were controlled at 291 K and 5.0×10^{-4} m^3/min. Then, we started the thermal energy storage with TBAAc hydrate. We set t = 0 when we started the circulation of the cooling water. The time evolution of T_{in},

T_{out} (temperature at inlet of the chiller) and T_{PCM} were measured every 10 s to evaluate the kinetic aspect of TBAAc hydrate as a thermal energy storage medium.

We performed such experiments in three systems. First, we did not add any external forces to the apparatus. Second, we used a rotating-impeller agitator to break adhesions between the hydrate and the coil. Third, we used the ultrasonic wave instead of mechanical agitation. In the system where the agitator was used, we also investigated the effect of rotation rate of the agitator by changing the rate to 100 rpm, 300 rpm and 600 rpm. Similarly, we investigated the effect of the frequency of ultrasonic waves on the system by changing the frequency. The frequency was set at either 28 kHz or 56 kHz.

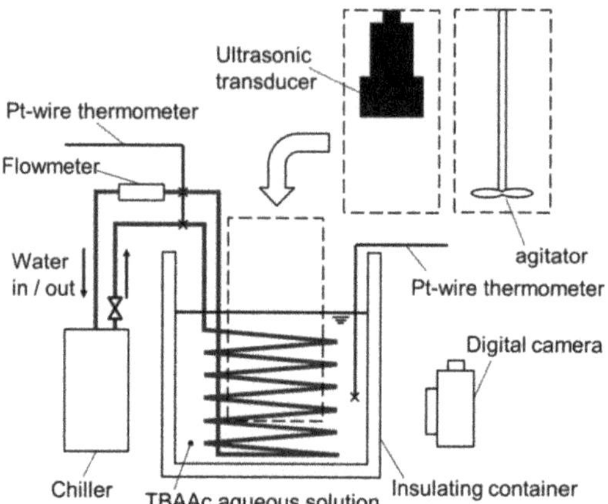

Figure 1. Schematic diagram of the apparatus.

3. Results and Discussion

3.1. The System without Any External Forces (Static System)

Initially, we reported the thermographs of Tetrabutylammonium Acrylate (TBAAc) hydrate obtained by a differential scanning calorimeter (DSC). Figure 2 shows the thermographs, which were measured at $w_{TBAAc} = 0.33$ and 0.36. A single peak was observed, and the existence of ice was not identified at $w_{TBAAc} = 0.33$.

As for the system without any external forces (static system), we performed thermal energy storage experiments. We measured T_{in}, T_{out}, and T_{PCM}. Figure 3 shows the time evolution of each temperature. Time zero indicates the time at which the circulating water is started to flow inside the coil. As shown in Figure 3, T_{PCM} decreased until t = 1.5 h and increased from t = 1.5 to t = 4 h. When T_{PCM} is increasing, TBAAc hydrate formation would occur because the T_{PCM} was 10 K lower than its equilibrium temperature, 291.5 K [35]. After the hydrate formation, the heat generation from the hydrate formation occurred and exceeded the endothermic of the heat exchanger. This was why T_{PCM} rose from t = 1.5 to t = 4 h. T_{PCM} moderately decreased again at t = 4 h when T_{PCM} was 288 K. At this moment, the formed TBAAc hydrate gradually grew around the heat exchanger as shown in Figure 4d. This adhesion would be the thermal resistance between the heat exchanger and thermal energy storage medium much lower than before. Then, the hydrate formation rate decreased due to the increased thermal resistance, and T_{PCM} decreased because the endothermic of the heat exchanger exceeds the heat generation from hydrate formation.

Figure 4a–d illustrates the optical observation of the time evolution of the hydrate formation. The hydrate was formed on the surface of the heat exchanger. Heat exchange occurs on the heat exchanger surface, which leads to hydrate formation occurring there. From t = 1.5 h to t = 2 h, the hydrate gradually grew and started covering the coil surface. The surface of the coil was fully covered by the hydrate around t = 4 h. After t = 7 h, optical changes could not be observed. In addition, we found that the TBAAc hydrate kept growing from the surface of the coil to the bulk of the solution. There was a 3 cm gap between the surface of the coil and the wall of the container. This discovery of the hydrate growth length can be a tip to design a static thermal energy storage system. Thermal energy storage rate, $\dot{q}$, and density, $\bar{q}$, were calculated by the formulas (1) and (2). What every character stands for is written in Table 2. On the right-hand side of (2), the thermal energy storage density due to sensible heat is subtracted from the thermal energy storage density, which is obtained by integrating the thermal energy storage rate.

$$\dot{q} = \rho_w \dot{V}_w c_w (T_{out} - T_{in}) / V_{PCM} \tag{1}$$

$$\bar{q} = \sum_{t=0}^{t} \dot{q} \, dt - \rho_{PCM} c_{PCM} (T_{PCM,\,0} - T_{PCM}) \tag{2}$$

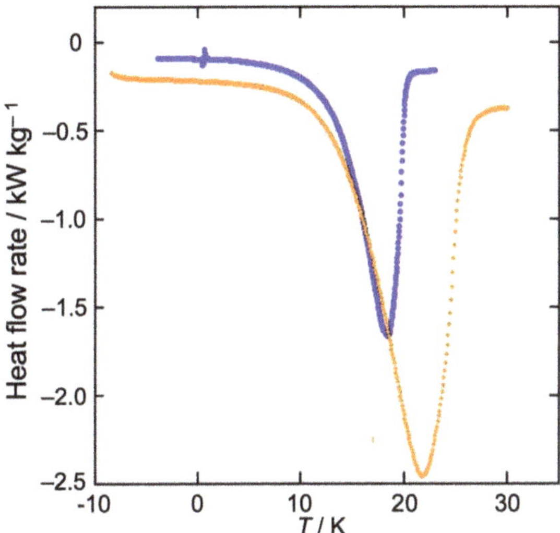

Figure 2. The DSC heating curves measured at w_{TBAAc} = 0.33 and 0.36. ●: w_{TBAAc} = 0.36, ◆: w_{TBAAc} = 0.33.

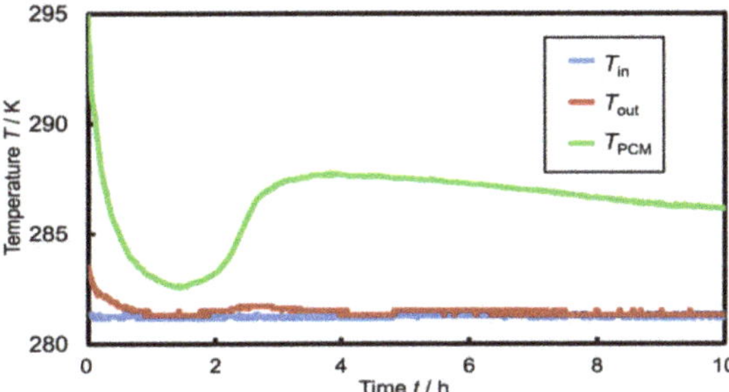

Figure 3. Time evolutions of the temperature at inlet of the chiller (T_{in}), outlet (T_{out}) of the chiller and thermal energy storage medium (T_{PCM}) in the static system.

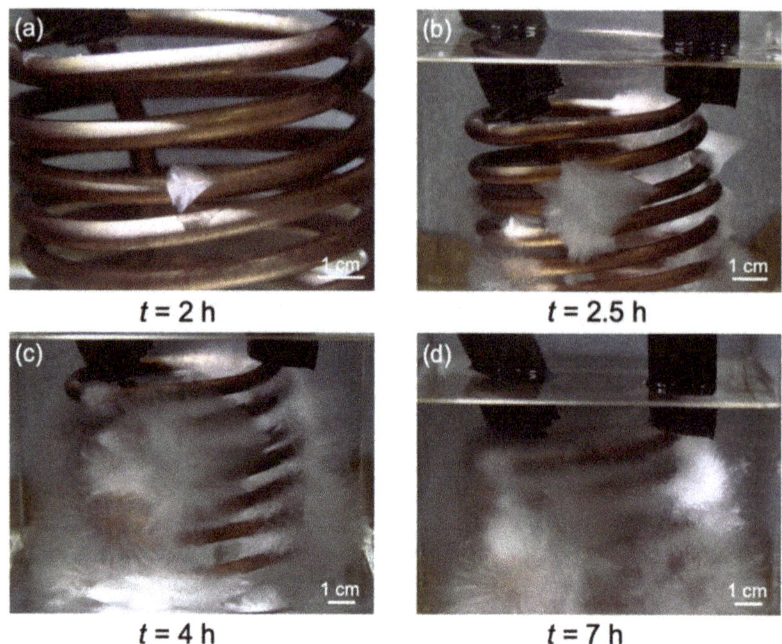

Figure 4. Crystal growth of TBAAc hydrate on the surface of the heat exchanger coil. (**a**–**d**) illustrate the optical measurement of the hydrate formation at t = 2, t = 2.5, t = 4 and t = 7, respectively.

Table 2. The value of each parameter in the calculation.

Parameter	
Density, ρ_w (Water)	999.8 kg/m^3
Volumetric flow rate, $\dot{V}_w$ (Water)	8.3×10^{-5} m^3/s
Specific heat at constant pressure, c_w (Water)	4.20 kJ/(kg K)
Density, ρ_{PCM} (PCM)	1.0 kg/m^3
Specific heat at constant pressure, c_{PCM} (PCM)	4.20 kJ/(kg K)

The obtained thermal energy storage density represents the heat storage density due to heat of hydrate formation and decomposition. The density of the PCM is assumed to be 1.0 kg m^{-3}. The value of each parameter used in the calculation is shown in Table 2. The physical property of water was referred to REFPROP ver. 9.1 [36].

Figure 5a illustrates the time evolution of the thermal energy storage rate in the static system. Since $\dot{q}$ is proportional to the gap between T_{in} and T_{out}, $\dot{q}$ varied the same way with T_{out}. Immediately after the start of the experiment, $\dot{q}$ decreased as T_{out} decreased, and as T_{out} approached T_{in}, $\dot{q}$ approached 0. After that, $\dot{q}$ increased at t = 1.5 h and reached around 9 kW/m^3. This is due to the increase in T_{out} caused by the heat generation from hydrate formation and growth, as described above. After t = 3 h, as T_{out} descended to be tangent to T_{in}, $\dot{q}$ descended to 0 with some fluctuations. Figure 5b illustrates the time evolution of thermal energy storage density. Since $\bar{q}$ indicates the thermal energy density by hydrate growth and decomposition, there is no change right after the experiments start. As the hydrate grew, thermal energy storage density increased. The rate of increase gradually became slower. The hydrate formation rate got slower due to the decreased heat conductivity around the heat exchanger coil. This is why the increasing rate of thermal energy storage density decreased. Thermal energy density reached 96 MJ/m^3 in 10 h.

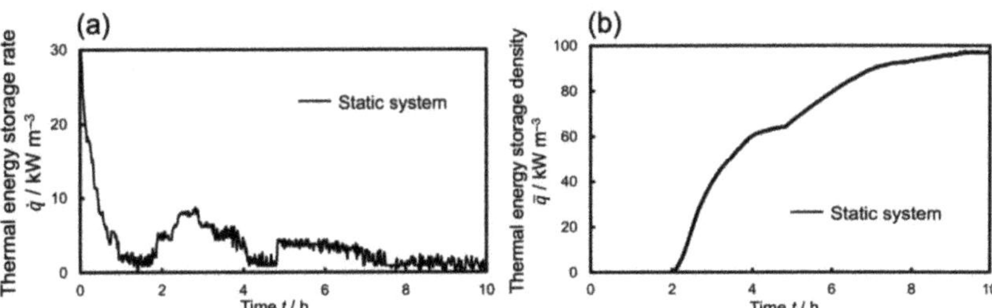

Figure 5. Time evolution of thermal energy storage rate and density in static system. (**a**,**b**) depict the thermal energy storage rate and density, respectively.

It is known that the solid phase fraction of ice thermal storage increases significantly with time up to around 30%, but the increasing rate decreases after about 40–50%. Therefore, it is favorable to design a thermal energy storage system that has a solid phase fraction of 40–50% [37]. For example, when the solid phase fraction is 45% in an ice thermal energy storage system, thermal energy storage density is approximately 140 MJ/m^3. Oyama et al. [33] reported that TBAB hydrate, which is the only commercially available hydrate PCM, has the heat for formation and decomposition of approximately 200 MJ/m^3. Assuming the TBAAc hydrate has 200 MJ/m^3 of heat in hydrate formation and decomposition, the thermal energy storage density corresponds to a solid phase fraction of 48%.

3.2. In the System with Mechanical Agitation

As written in 3.1., the surface of the heat exchanger coil was covered by hydrate. This hydrate adhesion increased thermal resistance. To solve this problem, mechanical agitation was used. In addition, to investigate the effect of the rotation rate, we changed the rate to 100 rpm, 300 rpm and 600 rpm. Figure 6a–c shows the time evolution of T_{in}, T_{out} and T_{PCM} in each rotation rate. Time zero was defined as the time at which the circulating water is started to be flown. In every rotation rate, T_{PCM} decreased twice after the experiment started and temporarily increased. In the agitated system, T_{PCM} increased earlier than the static system. As the rotation rate became quicker, T_{PCM} increased earlier. When T_{PCM} increased, the temperature was around 282 K, which is the same as the static system. This result indicates that the forced convection generated by the agitation increased the convective heat transfer coefficient, which resulted in faster cooling of the heat storage medium and thus faster nucleation. T_{PCM} increased rapidly more than the static system, which was almost vertically to around 288 K. This would be due to the increase in heat and mass transfer rates caused by the forced convection generated by the agitation, which led to the increase in the hydrate growth rate.

T_{PCM} changed from a rise to a fall around 288 K and then descended slowly. As the hydrate grew, the viscosity of the heat storage medium increased and the forced convection was weakened, which lowered the hydrate formation rate. T_{PCM} decreased because the hydrate growth rate became smaller, and the heat absorption by the heat exchanger exceeded the heat generated from hydrate formation.

Figure 7a–c shows the pictures of crystal growth in the system when the mechanical agitation rate was 300 rpm. At t = 0.6 h, numerous small crystals began to float in the aqueous solution. After t = 0.7 h, hydrate was formed in the entire aqueous solution. The hydrate nucleation occurred in the entire aqueous solution. This was because of the forced convection generated by the agitation, which stripped the hydrate from the heat exchange coil and suspended it in the aqueous solution. This convection increased the hydrate growth rate as well. Within a few minutes, the heat exchange coil was only faintly visible by the hydrate, as shown in Figure 7b. The hydrate formation has progressed and the viscosity increased as shown in Figure 7c, after which little change was observed.

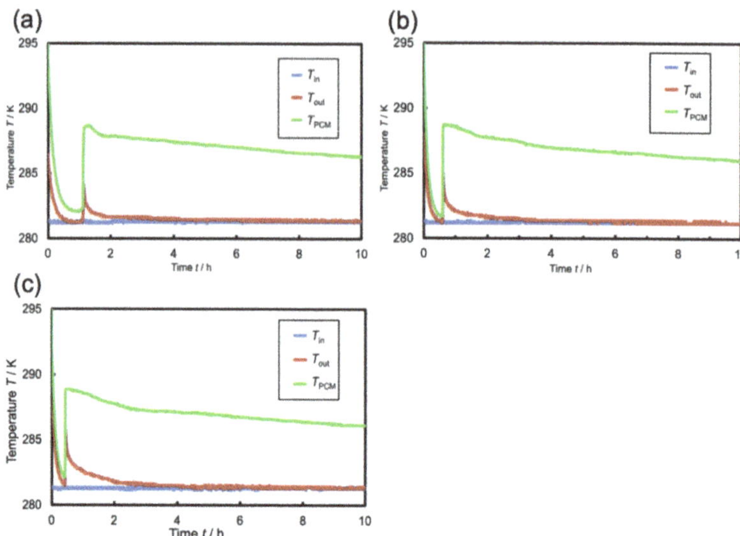

Figure 6. Time evolutions of the temperature at inlet of the chiller (T_{in}), outlet (T_{out}) of the chiller and thermal energy storage medium (T_{PCM}) in the system with the mechanical agitation. (**a–c**) depict each curve at the agitation rate of 100 rpm, 300 rpm and 600 rpm, respectively.

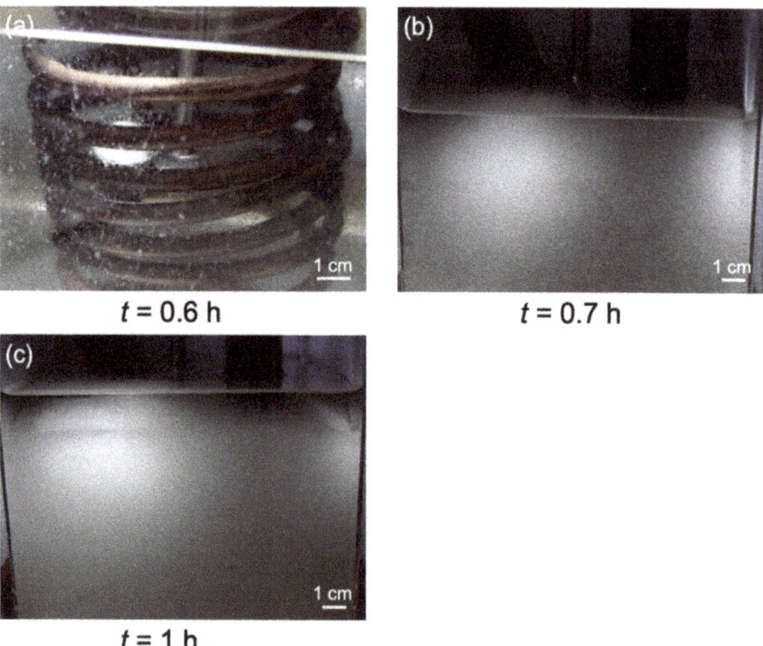

Figure 7. Optical observation changing in the container with mechanical agitation of 300 rpm. (**a–c**) are the pictures at 0.6 h, 0.7 h and 1 h after cooling water started to be circulated, re-spectively.

We visually observed that TBAAc hydrate formed and grew to the bulk of the solution. The gap between the heat transfer surface and the corner of the container was about 5 cm. In other words, when the cooling temperature was 10 K lower than the equilibrium temperature, TBAAc hydrate formed and grew at a distance of more than 5 cm from the heat transfer surface in the agitated system. The same tendency was observed at the rotation rate of 100 rpm and 600 rpm.

Figure 8a shows the time evolution of thermal energy storage density. To make a comparison with the static system, the time evolution of the static system is also shown in Figure 8a. Regarding the thermal energy storage rate, $\dot{q}$ is proportional to the difference between the inflow temperature T_{in} and the outflow temperature T_{out}. This was why $\dot{q}$ had almost the same trend as the T_{out} change in the static system. As T_{out} increased rapidly more than the static system, $\dot{q}$ increased rapidly more than the static system as well, rising almost vertically. The maximum thermal energy storage rate at that time was about 6 to 8 times faster than that of the static system. The larger the rotation rate of the agitation increased the value of $\dot{q}$. After a rapid increase, the change in $\dot{q}$ suddenly turned into a downward movement and gradually descended to 0. This was because the hydrate had grown, and the viscosity had increased, which resulted in less convection.

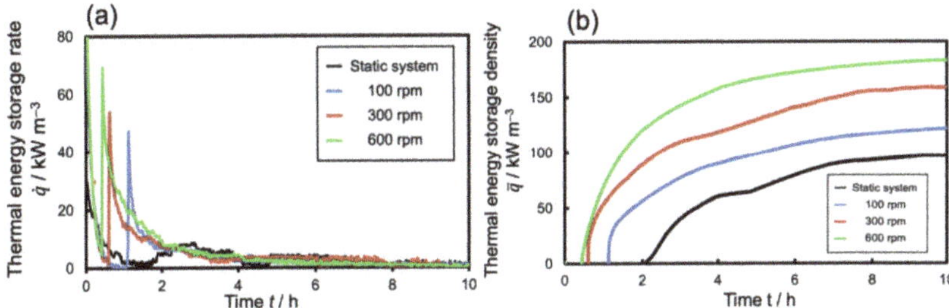

Figure 8. Time evolution of thermal energy storage rate and density in the agitated system (**a**,**b**) depict the thermal energy storage rate and density, respectively.

As shown in Figure 8b, $\bar{q}$ in the agitated system increased earlier than that in the static system, and the higher rotation rate of the agitation increased $\bar{q}$ earlier. This was because the temperature of the thermal energy storage medium was lowered quickly by forced convection, and thus hydrate nucleation also occurred quickly. The thermal energy storage density $\bar{q}$ was larger than that of the static system and increased by a higher rotation rate. This was because the forced convection caused by the agitation allowed hydrate to be formed in the entire bulk of the solution. In the agitated system with the rotation rates of 100 rpm, 300 rpm and 600 rpm, thermal energy density reached 121 MJ/m^3, 159 MJ/m^3 and 183 MJ/m^3 in 10 h, respectively. Similarly, assuming the TBAAc hydrate has 200 MJ/m^3 of heat in hydrate formation and decomposition, the thermal energy storage density corresponds to a solid phase fraction of 61%, 80% and 92%, respectively.

3.3. In the System Using an Ultrasonic Transducer

In this section, we discuss the results in the system with ultrasonic vibration. The inflow temperature T_{in}, outflow temperature T_{out} and thermal energy storage medium temperature T_{PCM} were measured. Figure 9a,b shows the time evolutions of T_{in}, T_{out} and T_{PCM} in the ultrasonic vibration system (28 kHz and 56 kHz). Time zero was defined as the time at which the circulating water was started to flow.

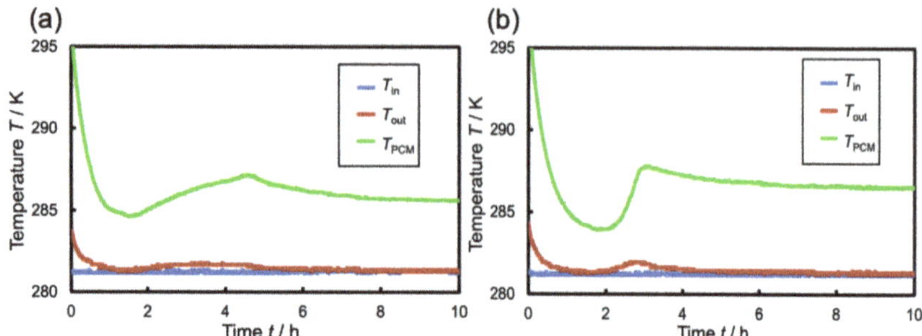

Figure 9. Time evolutions of the temperature at inlet of the chiller (T$_{in}$), outlet (T$_{out}$) of the chiller and thermal energy storage medium (T$_{PCM}$) in the system with ultrasonic vibration. (**a,b**) depict each curve at the frequency of 28 kHz and 56 kHz, respectively.

The slope of the T$_{PCM}$ first decrease was similar to that of the static system. However, when the slope changed from negative to positive, the T$_{PCM}$ was about 284 K, which was higher than that of the static and agitated system. This is because the hydrate nucleation occurred at a higher temperature than in the static system. In an ultrasonic vibration system, hydrate nucleation would occur at a higher temperature due to the effects of cavitation or acceleration of ultrasonic vibration.

The slope of the T$_{PCM}$ increase is not much different from that of the static system at 28 kHz and appears to be more gradual at 56 kHz. This was because of relatively high T$_{PCM}$, which results in a relatively slow hydrate growth rate and low heat of formation during hydrate nucleation. The increase in T$_{PCM}$ changed from a rise to a fall at around 288 K, and then T$_{PCM}$ slowly decreased. This was because the slope of the T$_{PCM}$ change turned from positive to negative because hydrate was generated and covered the heat exchange coil surface as the static system.

Figure 10a–c shows the pictures of the crystal growth during this system. As shown in Figure 10a–c, hydrate was formed on the heat transfer surface near the bottom of the container. As time went by, the formed hydrate piled up on the bottom of the container. Figure 11a shows time evolutions of thermal energy storage rate in the system using the ultrasonic vibration, which depends on the frequency of 28 kHz and 56 kHz. The thermal energy storage rate $\dot{q}$ in the static system is also shown for comparison. Time zero was defined as the time at which the circulating water is started to flow. Here, $\dot{q}$ was proportional to the difference between T$_{in}$ and T$_{out}$ and thus shows almost the same trend as T$_{out}$ change in the static system. In addition, the slope of change in the thermal energy storage rate was similar to that of the static system. However, as for the 28 kHz ultrasonic vibration, the maximum value of $\dot{q}$ after the increase was about 1.5 times larger than that of the static system. This was because the hydrate was detached from the heat exchange coil by cavitation, which increased the growth rate of the hydrate. On the other hand, the maximum value of $\dot{q}$ at 56 kHz was almost the same as that of the stationary system. Considering the size of hydrate adhesion, the cavitation effect is stronger at lower frequencies. This was why the maximum value of $\dot{q}$ at 56 kHz was larger than that of $\dot{q}$ at 28 kHz. After the peak of $\dot{q}$, the ultrasonic vibration system maintained a larger $\dot{q}$ for a relatively longer time than the static system. This was because the hydrate growth was continued for a relatively long time after the hydrate covered the heat exchange coil due to the effects of cavitation. Figure 11b shows the time evolution of the thermal energy storage density. The result of the static system is also shown in Figure 11b. Time zero was defined as the time at which the circulating water is started to flow. As shown in Figure 11b, $\bar{q}$ in the ultrasonic vibrated system started increasing a bit earlier than the static system. This was because the cavitation enabled the hydrate to be formed at a higher temperature. The slope of $\bar{q}$ change in the ultrasonic vibration system is almost the same as with the static system. However, the bigger slope could be kept longer than the static system. This was also because the hydrate growth could be kept by the effect of the cavitation. In the ultrasonic vibration system with a frequency of 28 kHz and 56 kHz, thermal energy storage density reached 126 MJ/m^3, 111 MJ/m^3 in 10 h, respectively. Similarly, assuming the TBAAc hydrate has 200 MJ/m^3 of heat in hydrate formation and decomposition, the thermal energy storage density corresponds to a solid phase fraction of 63% and 55%, respectively.

Figure 10. Optical observation changing in the container with ultrasonic vibration of 28 kHz. (**a–d**) are the pictures at 2 h, 2.5 h, 3 h and 5 h after cooling water started to be circulated, respectively.

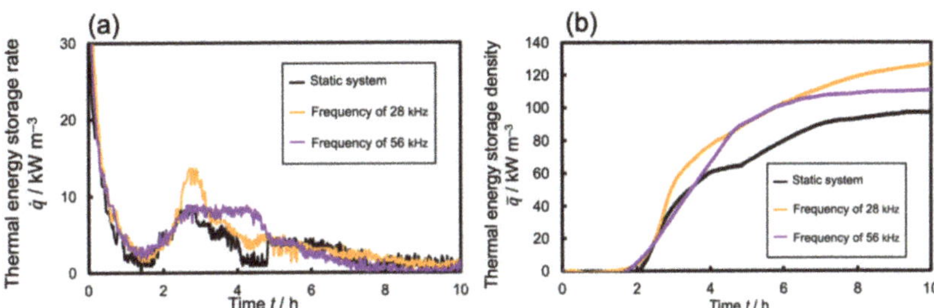

Figure 11. Time evolutions of the thermal energy storage rate and density in the system using an ultrasonic vibration. (**a,b**) depict the thermal energy storage rate and density, respectively.

3.4. Comparison in All Systems

Figure 12a shows a comparison in the time evolution of the thermal energy storage rate in all systems. In the agitated system, the thermal energy storage rate, $\dot{q}$, increased earlier than the other systems; $\dot{q}$ increased at 1 h, 0.6 and 0.4 h for a rotation rate of 100 rpm, 300 rpm and 600 rpm, respectively. The forced convection generated by the agitation increased the convective heat transfer coefficient. This was why the temperature of the thermal energy storage medium dropped relatively earlier and hydrate formation occurred earlier, resulting in the early increase of $\dot{q}$. In addition, this forced convection caused rapid hydrate formation in the entire container, which released huge heat in a short period. This was why $\dot{q}$ increased vertically. As the rotation rate increased, $\dot{q}$ increased more.

The maximum values of $\dot{q}$ after the vertical increase were 47 kW m^{-3}, 54 kW m^{-3} and 69 kW m^{-3} for a rotation rate of 100 rpm, 300 rpm and 600 rpm, respectively. These values are 5 to 8 times larger than that of a static system.

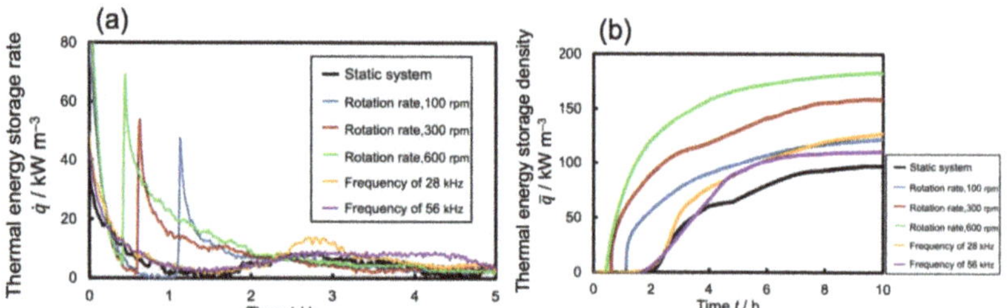

Figure 12. Time evolutions of thermal energy storage rate and density in all systems. (**a,b**) depict the thermal energy storage rate and density, respectively.

In the ultrasonic vibration system, the maximum of $\dot{q}$ was 14 kW m^{-3}, which is about 1.5 times larger than that of the static system, when the frequency was 28 kHz. This was because cavitation detached the adhesion between hydrate and heat exchanger coil, which could be the thermal resistance. As for the system with 56 kHz frequency, the maximum values of $\dot{q}$ were almost the same as in the static system. This was because the effect of the cavitation was not as big as 28 kHz [38]. In summary, the agitation increased the rate of heat and mass transfer by creating forced convection, thereby shortening the hydrate formation time and increasing the hydrate growth rate. On the other hand, ultrasonic vibration kept the hydrate growth longer by the effect of cavitation.

Figure 12b shows a comparison in the time evolution of thermal energy storage density, $\bar{q}$, in all systems. In the agitated system, $\bar{q}$ increased earlier and quicker than the other systems. This was because T$_{\text{PCM}}$ decreased quicker by the forced convection, which caused the quicker hydrate formation and growth for the entire container. In the ultrasonic vibration system, the $\bar{q}$ increase appeared to be relatively longer than the other systems. This was because the cavitation enabled hydrate to grow after hydrate covered the surface of the heat exchanger coil. The static system, the agitated system (rotation rate of 100 rpm, 300 rpm and 600 rpm), and the ultrasonic vibrated system (frequency of 28 kHz and 56 kHz) had the thermal energy storage density of 96 MJ/m^3, 121 MJ/m^3, 159 MJ/m^3, 183 MJ/m^3, 126 MJ/m^3, and 111 MJ/m^3, respectively. Assuming TBAAc hydrate has the heat for formation and decomposition of approximately 200 MJ/m^3, each solid phase fraction is 48%, 61%, 80%, 92%, 63%, and 55%, respectively.

3.5. For Industrial Utilizing of TBAAc Hydrate as PCM

As written in Section 3.1., a range of the practical solid-phase fraction is from 40 to 50% [37]. The thermal energy storage density of ice is 140 MJ m^{-3} when the solid phase fraction is 45%. As for TBAB hydrate, it has a thermal energy storage density of 100 MJ/m^3 when the solid phase fraction is 50% [33]. We regarded these values as a target thermal energy storage density. Given the thermal energy storage during nighttime of 10 h, the required time for thermal energy storage should be shortened. This was why the time to reach the target values was compared to each other system. In addition, the amount of the used electric power was estimated for more practical evaluation.

Table 3 shows the time that was needed to achieve the target thermal energy storage density. As for the target of 140 MJ/m^3, it was achieved only when the rotation rates of agitation were 300 rpm and 600 rpm. When the rotation rates were 300 rpm or 600 rpm, it was 5.9 h or 2.9 h to achieve the target, respectively. Concerning the consumed electric power to achieve the target, it was 58 MJ/m^3 and 29 MJ/m^3, respectively. As for the target of 100 MJ/m^3, it was achieved in all systems where an external force was applied. When the rotation rates were 300 rpm and 600 rpm, it took about a third of the time of the ultrasonic vibration system. However, the consumed electric power by the external force to achieve the target was 45 MJ/m^3, 24 MJ/m^3, 15 MJ/m^3 for the agitation system (100 rpm, 300 rpm, 600 rpm), and 5 MJ/m^3 for the ultrasonic vibration system at both frequencies (28 kHz, 56 kHz), respectively. Therefore, when the target of thermal energy storage density is 100 MJ/m^3, the agitation system (300 rpm, 600 rpm) is more suitable for storing thermal energy in a short time, and the ultrasonic vibration system is more suitable as an external force for reducing the amount of electricity used.

Table 3. The time to achieve the target thermal energy storage density.

Target	Static System	Agitated System			Ultrasonic Vibrated System	
		100 rpm	300 rpm	600 rpm	28 kHz	56 kHz
100 MJ/m^3 Achieved time/h	-	5.1	2.4	1.5	5.8	5.8
140 MJ/m^3 Achieved time/h	-	-	5.9	2.9	-	-

In this study, the thermal energy storage density in the agitated system was greatly improved by adding agitation. On the other hand, the amount of electricity used for agitation became larger than other systems. The agitation intensity was almost the same as that in a general low-viscosity agitation. However, the agitation Reynolds number was 25,000 at 600 rpm. This was due to the fact that the impeller power number of the three propeller blades was 1, which was smaller than that of other agitation blades [39]. Therefore, even if the agitation intensity was a general value, the agitation Reynolds number became larger, and energy consumption also became larger. It will be possible to optimize the rotations and electric power consumption by selecting a more suitable agitator at the similar experimental apparatus. The thermal energy storage density in the ultrasonic vibration system was not as large as that in the agitation system. The reason for this result is that it was difficult to resonate the ultrasonic transducer. To resonate the ultrasonic transducer, it is necessary to provide a natural frequency, but this frequency varies slightly depending on the environment in which the ultrasonic transducer is placed, for example, the materials in contact. Also, the natural frequency is also expected to change during the transformation of the aqueous solution into hydrate. Therefore, it is possible that the ultrasonic transducer was not able to resonate properly in this experimental system. A possible solution to solve this problem is to use a device that automatically follows the resonant frequency and keeps giving the resonant frequency to the transducer. Since the power consumption for ultrasonic vibration is about one-tenth of the power used for agitation, it is thought that the ultrasonic vibration system could be the most favorable way by solving the above problems.

3.6. The Evaluation of the Heat Exchanger

The thermal energy storage rate by the heat exchanger can be calculated as in Equation (3), where U is the heat transfer coefficient, A is the surface area of the heat exchange coil, and V_{PCM} is the volume of the thermal energy storage medium. The logarithmic mean temperature difference, ΔT_{lm}, is calculated by Equation (4). As can be seen from Equation (4), the logarithmic mean temperature difference, ΔT_{lm}, is the average of the temperature difference between the thermal energy storage medium and the cooling water using the logarithm. This logarithmic mean temperature difference was substituted into Equation (3) to calculate the heat transfer coefficient for the experiments under each condition. The heat transfer coefficient was calculated to be about 10^4–10^5 W m^{-2} K. The A/V_{PCM} in this experiment was 15 m^2/m^3, and the UA/V_{PCM} value was calculated to be around 10^3–10^4 W m^{-3} K. For instance, a plate heat exchanger has a UA/V_{PCM} that is similar to this value [40]. As the UA/V increases, the amount of heat exchange will also increase.

It realizes a higher thermal energy storage density. In this study, the UA/V of the heat exchanger was comparable to that of other common heat exchangers [40]. This result demonstrates that the comparable thermal energy storage density in this study can be obtained using other common heat exchangers.

For the practical use of thermal energy storage using TBAAc hydrate on a larger scale, it is also expected that the comparable thermal energy storage density in this study would be available by using a heat exchanger which has the same level UA/V.

$$\dot{q} = UA\Delta T_{\mathrm{lm}}/V_{PCM} \tag{3}$$

$$\Delta T_{\mathrm{lm}} \frac{T_{out} - T_{in}}{ln\frac{T_{PCM}-T_{in}}{T_{PCM}-T_{out}}} \tag{4}$$

4. Conclusions

Kinetic characteristics of thermal energy storage using tetrabutylammonium acrylate (TBAAc) hydrate were practically and experimentally evaluated for practical use as Phase Change Materials (PCMs). During the experiments, the adhesion of hydrate on the surface of the heat exchanger coil was found, which increased the thermal resistance between the heat exchanger and thermal energy storage medium. This was why, as the external forces, we added a mechanical agitation (rotation rate of 100 rpm, 300 rpm, and 600 rpm) and ultrasonic vibration (frequency of 28 kHz and 56 kHz) on each system. It was revealed that the external forces improved the thermal energy storage kinetic characteristics. When the agitation rate was 600 rpm, the system achieved thermal energy storage of 140 MJ/m^3 in 2.9 h. This value is comparable to the ideal performance of ice thermal energy storage when its solid phase fraction is 45%. The energy consumption for agitation was 28 MJ/m^3 to achieve this value. The thermal energy storage of 100 MJ/m^3, which is a TBAAc solid phase fraction of 50%, was achieved in 1.5 h and 5.8 h with mechanical agitation (600 rpm) and ultrasonic vibration (28 kHz), respectively. The energy consumptions to achieve this target value were 15 MJ/m^3 and 5.2 MJ/m^3, respectively. In summary, the agitation increased the rate of heat and mass transfer by creating forced convection, thereby shortening the hydrate formation time and increasing the growth rate of the hydrate. In addition, ultrasonic vibration could keep the hydrate growth time longer by the effect of cavitation.

In this system, the UA/V (U: thermal transfer coefficient, A: surface area of the heat exchange coil, V: volume of the thermal energy storage medium) was approximately 1.0×10^{-4}–10^{-3} W m^{-3} K. This UA/V of the heat exchanger was comparable to that of other common heat exchangers. Even if the system would be scaled up and different from this study, the same level of thermal energy storage as this study could be obtained by using a heat exchanger of which UA/V is comparable to this study. As for the ultrasonic vibration system, thermal energy storage density was one-tenth of the agitated system. A device that enables an oscillator to resonate and break the hydrate adhesion is necessary to increase the thermal energy storage density. Then, that system would be most favorable in terms of the storing performance and energy power consumption.

Author Contributions: Conceptualization, H.K., H.T. and R.O.; methodology, H.T., S.I., H.N. and R.O.; formal analysis, H.T., S.I., H.N. and R.O.; investigation, H.K., H.T., S.I., H.N. and R.O.; resources, H.T. and R.O.; data curation, H.K., H.T., S.I., H.N. and R.O.; writing—original draft preparation, H.K. and R.O.; writing—review and editing, H.K. and R.O.; visualization, H.T. and R.O.; supervision, R.O.; project administration, R.O.; funding acquisition, R.O. All authors have read and agreed to the published version of the manuscript.

Funding: This work has been supported by a Keirin-racing-based research promotion fund from the JKA Foundation (Grant Number: 2020M-195) and a part of Low Carbon Technology Research and Development Program for "Practical Study on Energy Management to Reduce CO$_2$ emissions from University Campuses" from the Ministry of the Environment, Japan.

Institutional Review Board Statement: Not applicable.

Informed Consent Statement: Not applicable.

Data Availability Statement: Not applicable.

Acknowledgments: The authors would like to thank Haruki Sato, professor emeritus and Masao Takeuchi, Keio University, for the encouragement on this work.

Conflicts of Interest: The authors declare no conflict of interest.

References

1. IEA. *World Energy Outlook*; International Energy Agency (IEA): Paris, France, 2019.
2. Fei, F.; Qi, Q.; Liu, A.; Kusiak, A.A. Data-driven smart manufacturing. *J. Manuf. Syst.* **2018**, *48*, 157–169.
3. Król, A. The Application of the Artificial Intelligence Methods for Planning of the Development of the Transportation Network. *Transp. Res. Procedia* **2016**, *14*, 4532–4541. [CrossRef]
4. Masanet, E.; Shehabi, A.; Lei, N.; Smith, S.; Koomey, J. Recalibrating global data center energy-use estimates. *Science* **2020**, *367*, 984–986. [CrossRef] [PubMed]
5. Zimmermann, S.; Meijer, I.; Tiwari, M.K.; Paredes, S.; Michel, B.; Poulikakos, D. Aquasar: A hot water cooled data center with direct energy reuse. *Energy* **2012**, *43*, 237–245. [CrossRef]
6. Beghi, A.; Cecchinato, L.; Mana, G.D.; Lionello, M.; Rampazzo, M.; Sisti, E. Modelling and control of a free cooling system for Data Centers. *Energy Procedia* **2017**, *140*, 447–457. [CrossRef]
7. Ko, J.-S.; Huh, J.-H.; Kim, J.-C. Improvement of Energy Efficiency and Control Performance of Cooling System Fan Applied to Industry 4.0 Data Center. *Electronics* **2019**, *8*, 582. [CrossRef]
8. Khalaj, A.H.; Halgamuge, S.K. A Review on efficient thermal management of air- and liquid-cooled data centers: From chip to the cooling system. *Appl. Energy* **2017**, *205*, 1165–1188. [CrossRef]
9. Stephan, K.; Laesecke, A. The thermal conductivity of fluid air. *J. Phys. Chem. Ref. Data* **1985**, *14*, 227–234. [CrossRef]
10. Pielichowska, K.; Pielichowski, K. Phase change materials for thermal energy storage. *Prog. Mater. Sci.* **2014**, *65*, 67–123. [CrossRef]
11. Veerakumar, C.; Sreekumar, A. Phase change material based cold thermal energy storage: Materials, techniques and applications—A review. *Int. J. Refrig.* **2015**, *67*, 271–289. [CrossRef]
12. Xie, N.; Huang, Z.; Luo, Z.; Gao, X.; Fang, Y.; Zhang, Z. Inorganic Salt Hydrate for Thermal Energy Storage. *Appl. Sci.* **2017**, *7*, 1317. [CrossRef]
13. Porteiro, J.; Míguez, J.L.; Crespo, B.; Lara, J.; Pousada, J.M. On the Behavior of Different PCMs in a Hot Water Storage Tank against Thermal Demands. *Materials* **2016**, *9*, 213. [CrossRef] [PubMed]
14. Ndukwu, M.C.; Bennamoun, L.; Simo-Tagne, M. Reviewing the Exergy Analysis of Solar Thermal Systems Integrated with Phase Change Materials. *Energies* **2021**, *14*, 724. [CrossRef]
15. Wong-Pinto, L.-S.; Milan, Y.; Ushak, S. Progress on use of nanoparticles in salt hydrates as phase change materials. *Renew. Sustain. Energy Rev.* **2020**, *122*, 109727. [CrossRef]
16. Mitran, R.-A.; Ionita, S.; Lincu, D.; Berger, D.; Matei, C. A Review of Composite Phase Change Materials Based on Porous Silica Nanomaterials for Latent Heat Storage Applications. *Molecules* **2021**, *26*, 241. [CrossRef]
17. Zalba, B.; Marín, J.; Cabeza, L.; Mehling, H. Review on thermal energy storage with phase change: Materials, heat transfer analysis and applications. *Appl. Therm. Eng.* **2003**, *23*, 251–283. [CrossRef]
18. ASHRAE Technical Committee. *ASHRAE TC9.9 Data Center Power Equipment Thermal Guidelines and Best Practices*; ASHRAE: Peachtree Corners, GA, USA, 2016.
19. Arai, Y.; Yamauchi, Y.; Tokutomi, H.; Endo, F.; Hotta, A.; Alavi, S.; Ohmura, R. Thermophysical property measurements of tetrabutylphosphonium acetate (TBPAce) ionic semiclathrate hydrate as thermal energy storage medium for general air conditioning systems. *Int. J. Refrig.* **2018**, *88*, 102–107. [CrossRef]
20. Koyama, R.; Hotta, A.; Ohmura, R. Equilibrium temperature and dissociation heat of tetrabutylphosphonium acrylate (TBPAc) ionic semi-clathrate hydrate as a medium for the hydrate-based thermal energy storage system. *J. Chem. Thermodyn.* **2020**, *144*, 106088. [CrossRef]
21. Miyamoto, T.; Koyama, R.; Kurokawa, N.; Hotta, A.; Alavi, S.; Ohmura, R. Thermophysical property measurements of tetra-butylphosphonium oxalate (TBPOx) ionic semiclathrate hydrates as a media for the thermal energy storage system. *Front. Chem.* **2020**, *8*, 547. [CrossRef]
22. Koyama, R.; Arai, Y.; Yamauchi, Y.; Takeya, S.; Endo, F.; Hotta, A.; Ohmura, R. Thermophysical Properties of Trimethylolethane (TME) Hydrate as Phase Change Material for Cooling Lithium-ion Battery in Electric Vehicle. *J. Power Sources* **2019**, *427*, 70–76. [CrossRef]
23. Nakane, R.; Shimosato, Y.; Gima, E.; Ohmura, R.; Senaha, I.; Yasuda, K. Phase equilibrium condition measurements inn carbo dioxide hydrate forming system coexisting with seawater. *J. Chem. Thermodyn.* **2021**, *152*, 106276. [CrossRef]
24. Alavi, S.; Ohmura, R. Understanding decomposition and encapsulation energies of structure I and II clathrate hydrates. *J. Chem. Phys.* **2016**, *145*, 154708. [CrossRef] [PubMed]
25. Horii, S.; Ohmura, R. Continuous separation pf CO_2 from a $H_2 + CO_2$ gas mixture using clathrate hydrate. *Appl. Energy* **2018**, *225*, 78–84. [CrossRef]

26. Kiyokawa, H.; Horii, S.; Alavi, S.; Ohmura, R. Improvement of continuous hydrate-based CO_2 separation by forming structure II hydrate in the system of H_2 + CO_2 + H_2O + Tetrahydropyran (THP). *Fuel* **2020**, *278*, 118330. [CrossRef]
27. Hatsugai, T.; Nakayama, R.; Tomura, S.; Akiyoshi, R.; Nishitsuka, S.; Nakamura, R.; Takeya, S.; Ohmura, R. Development and Continuous Operation of a Bench-Scale System for the Production of O_3 + O_2 + CO_2 Hydrates. *Chem. Eng. Technol.* **2020**, *43*, 2307–2314. [CrossRef]
28. Nagashima, H.D.; Alavi, S.; Ohmura, R. Preservation of carbon dioxide clathrate hydrate in the presence of fructose or glucose and absence of sugars under freezer conditions. *J. Ind. Eng. Chem.* **2017**, *54*, 332–340. [CrossRef]
29. Kondo, Y.; Alavi, S.; Takeya, S.; Ohmura, R. Characterization of the Clathrate Hydrate Formed with Fluoromethane and Pinacolone: The Thermodynamic Stability and Volumetric Behavior of the Structure H Binary Hydrate. *J. Phys. Chem. B* **2021**, *125*, 328–337. [CrossRef]
30. Yamauchi, Y.; Yamasaki, T.; Endo, F.; Hotta, A.; Ohmura, R. Thermodynamic Properties of Ionic Semiclathrate Hydrate Formed with Tetrabutylammonium Propionate. *Chem. Eng. Technol.* **2017**, *40*, 1810–1816. [CrossRef]
31. Yamauchi, Y.; Arai, Y.; Yamasaki, T.; Endo, F.; Hotta, A.; Ohmura, R. Phase equilibrium temperature and dissociation heat of ionic semiclathrate hydrate formed with tetrabutylammonium butyrate. *Fluid Phase Equilib.* **2017**, *441*, 54–58. [CrossRef]
32. Sato, K.; Tokutomi, H.; Ohmura, R. Phase equilibrium of ionic semi-clathrate hydrates formed with tetrabutylammonium bromide and tetrabutylammonium chloride. *Fluid Phase Equilib.* **2013**, *337*, 115–118. [CrossRef]
33. Oyama, H.; Shimada, W.; Ebinuma, T.; Kamata, Y.; Takeya, S.; Uchida, T.; Nagao, J.; Narita, H. Phase diagram, latent heat, and specific heat of TBAB semi-clathrate hydrate crystals. *Fluid Phase Equilib.* **2005**, *234*, 131–135. [CrossRef]
34. Fujiura, K.; Nakamoto, Y.; Taguchi, Y.; Ohmura, R.; Nagasaka, Y. Thermal conductivity measurements of semi-clathrate hydrates and aqueous solutions of tetrabutylammonium bromide (TBAB) and tetrabutylammonium chloride (TBAC) by the transient hot-wire using parylene-coated probe. *Fluid Phase Equilib.* **2016**, *413*, 129–136. [CrossRef]
35. Sakamoto, H.; Sato, K.; Shiraiwa, K.; Takeya, S.; Nakajima, M.; Ohmura, R. Synthesis, Characterization and thermal-property measurements of ionic semi-clathrate hydrates formed with tetrabutylphosphonium chloride and tetrabutylammonium acrylate. *RSC Adv.* **2011**, *1*, 315–322. [CrossRef]
36. Lemmon, E.W.; McLinden, M.O. Huber M.L. NIST Reference Fluid Thermodynamic and Transport Properties (REFPROP). *J. Res. Natl. Inst. Stand. Technol.* **2013**.
37. Nakajima, M.; Hirata, A. *High-Efficient Thermal Energy Storage Technique with a Clathrate Hydrate; IHI Technical Report*; IHI: Tokyo, Japan, 2009; Volume 49, pp. 210–218.
38. Iida, Y. Sono Process no Hanashi—Chemical Engineering Applications of Ultrasonic (SCIENCE AND TECHNOLOGY). *Nikkan Kogyo Shimbun* **2006**, *7*, 7–17.
39. Hashimoto, K. *Industrial Reaction Equipment: Selection, Design, and Examples*; Baihukan: Tokyo, Japan, 1984; ISBN 9784563041618.
40. The Japan Society of Mechanical Engineers. *JSME Data Book: Heat Transfer Materials*; The Japan Society of Mechanical Engineers: Tokyo, Japan, 1986; ISBN 9784888981842.

Article

Exploitation of Enzymes for the Production of Biofuels: Electrochemical Determination of Kinetic Parameters of LPMOs

Dimitrios Zouraris [1], Anthi Karnaouri [2], Raphaela Xydou [1], Evangelos Topakas [2] and Antonis Karantonis [1,*]

[1] Laboratory of Physical Chemistry and Applied Electrochemistry, School of Chemical Engineering, National Technical University of Athens, Zografou, 15780 Athens, Greece; dimzouraris@mail.ntua.gr (D.Z.); ch15611@mail.ntua.gr (R.X.)

[2] Biotechnology Laboratory, School of Chemical Engineering, National Technical University of Athens, Zografou, 15780 Athens, Greece; akarnaouri@chemeng.ntua.gr (A.K.); vtopakas@chemeng.ntua.gr (E.T.)

* Correspondence: antkar@central.ntua.gr

Abstract: Lytic polysaccharide monooxygenases (LPMOs) consist of a class of enzymes that boost the release of oxidised products from plant biomass, in an approach that is more eco-friendly than the traditional ones, employing harsh chemicals. Since LPMOs are redox enzymes, they could possibly be exploited by immobilisation on electrode surfaces. Such an approach requires knowledge of kinetic and thermodynamic information for the interaction of the enzyme with the electrode surface. In this work, a novel methodology is applied for the determination of such parameters for an LPMO from the filamentous fungus *Thermothelomyces thermophila*, *Mt*LPMO9H.

Keywords: LPMO; Fourier Transform ac Voltammetry (FTacV); cyclic voltammetry; Direct Electron Transfer (DET)

Citation: Zouraris, D.; Karnaouri, A.; Xydou, R.; Topakas, E.; Karantonis, A. Exploitation of Enzymes for the Production of Biofuels: Electrochemical Determination of Kinetic Parameters of LPMOs. *Appl. Sci.* **2021**, *11*, 4715. https://doi.org/10.3390/app11114715

Received: 23 April 2021
Accepted: 18 May 2021
Published: 21 May 2021

Publisher's Note: MDPI stays neutral with regard to jurisdictional claims in published maps and institutional affiliations.

1. Introduction

The exploitation of lignocellulosic biomass, such as wood and agricultural residues, towards the production of fermentable sugars for the production of second-generation biofuels in an environmentally friendly way, requires of enzymatic degradation. This step has low energy requirements, selectivity, mild processing environment and no need of harsh chemicals, in contrast to classical chemical methods, and it is catalyzed by a combination of hydrolytic and oxidative enzymes [1,2]. Regarding the oxidation reactions, performing these steps in an electrochemical reactor, i.e., electroenzymatic synthesis of sugars, is a promising approach for the production of such chemicals. In principle, when electrolysis is combined with enzyme reactions, a redox enzyme is maintained in its active state through an electrode process. This can be achieved by oxidation of or a reduction in a cofactor or a mediator on the electrode surface, or by direct electron transfer to the enzyme. Therefore, applying enzyme immobilization onto the electrode and combining bioelectrocatalysis with traditional enzymatic hydrolysis offer perspectives for developing novel processes for the production of biofuels and other value-added materials from lignocellulose.

Lytic polysaccharide monooxygenases (LPMOs) are copper-dependent enzymes that have the capacity to cleave polysaccharides via an oxidative mechanism [3]. They are classified as Auxiliary Activities (AAs) in the Carbohydrate Active enZyme database (CAZy). More specifically, they are divided into seven families (AA9–11 and AA13–16) based on their sequence similarities, with the majority of those that are studied thoroughly belonging to the AA9 and AA10 families [4]. LPMOs act synergistically with other biomass-degrading enzymes and significantly boost the hydrolytic performance of cellulase and hemicellulase cocktails, primarily by improving the accessibility of the enzymes to the substrate [5,6].

Until recently, the main theory regarding the mechanism of action of LPMOs involved the introduction of a single oxygen atom from molecular oxygen into the substrate, while taking electrons from an external electron donor for their action [7]. More recently, Bissaro et al., while examining an LPMO belonging to the AA10 family, supported the idea that the chemically relevant co-substrate of LPMO catalyzed oxidation is H_2O_2 [8]. Regardless of the mechanism of the oxidative cleavage of the polysaccharidic substrate, in both cases, the reaction is initiated when the active center of the enzyme takes an electron from an external electron donor. The nature of the donors used in laboratory experimental setups include ascorbic acid, cysteine and reduced glutathione, but also a wide range of plant-derived phenols [9,10], lignin compounds [11] and photocatalytic and enzymatic systems [12]. The reported mechanisms for the initial reduction in LPMOs do not implicate any chemical steps other than a one-step, one-electron reaction.

One interesting point regarding these enzymes, which could possibly be an advantage for their electrochemical exploitation, is that their copper-containing active center is not "buried" inside their structure but can be found on a solvent-exposed, flat surface [13]. This could facilitate their use as electrode coatings, by directly providing them with electrons from an electrode surface, while immobilized on the electrode surface. Other than their study, this characteristic could lead to the exploitation of the bioelectrocatalytic action of LPMOs towards the production of oxidised products (oligosaccharides and glucans) from biomass, in concert with hydrolytic enzymes, in the absence of any external reducing agent. This means that by providing electrons directly from an electrode surface to an immobilised LPMO, the enzyme could be utilised in bioconversions through bioelectrocatalysis focusing on product life cycles. The developed bioconversions will be carried out in predominantly aqueous media using enzymes, and therefore be characterised by a limited use of toxic reagents or solvents, requiring only an ambient temperature, in the absence of any reducing agent [14].

In order to test if the above task is feasible, a parameter that should be mentioned is the standard electron transfer rate constant (k^0) between the immobilized electrode and the enzyme. This electron transfer rate should be rather fast, so that the electrocatalytic current measured represents the rate of substrate turnover at the enzyme active site [15]. Among the electroanalytical methods, a technique that is very promising for the study of immobilised LPMOs is that of large amplitude fast Fourier transform alternating current voltammetry (FTacV), with which one can study both free and immobilised species, eliminating capacitance currents, a problem common in immobilised-on-electrode species [16–19]. In the past, this technique has been employed for the determination of the formal potential of immobilised LPMOs [20], as well as for the study of the direct electron transfer of multi-copper laccase like oxidases [6] and the interaction with epinephrine as a substrate [21]. Recently, a methodology for the extraction of the transfer coefficient a and the k^0 for immobilised species has been proposed, but has yet to be applied experimentally [17]. In this work, this particular methology shall be employed for the extraction of the k^0 an LPMO from the filamentous fungus *Thermothelomyces thermophila*, *Mt*LPMO9H, and the calulated value of k^0 shall be verified by numerical simulation. To our knowledge, k^0 values for the interaction of an immobilised LPMO with an electrode surface is reported for the first time.

2. Materials and Methods

Voltammetric experiments were performed in a single-compartment three-electrode cell consisting of a working electrode, a Pt wire as a counter electrode, and a Ag|AgCl, KCl sat. reference electrode (+0.197 V vs. NHE). The aqueous solutions of about 5 mL consisted of 0.5 M acetate buffer (pH 6) as a supporting electrolyte.

The working electrode consisted of a cobalt functionalised multi-walled carbon nanotubes modified glassy carbon (GC) electrode, of 3 mm diameter. The immobilisation matrix was prepared as follows: A total of 1 g of commercial functionalised multi-wall carbon nanotubes (MWCNTs) (Hongwu Material, item No: C933-MC, Guangzhou, China) were added in a 150 mL solution with 20 % *v/v*. The suspension was stirred for 20 min at 120 °C for further functionalization, until the liquid phase was evaporated. The solution was left to dry overnight at 90 °C until dry powder was collected. The powder was washed, then centrifuged 5 times with deionised water until the supernatant had a pH value of $\simeq$7.0. This step ensured the removal of acetic acid excess. The obtained functionalised commercial MWCNTs (f-MWCNT) were left to dry at 90 °C. Subsequently, the f-MWCNTs were mixed with 150 mL deionised water and 0.4227 g of $(CH_3COO)_2Co \cdot 4H_2O$ (MERCK) leading to a loading 10 atom% Co on f-MWCNTs, which was then subjected to sonication at a low frequency (20 kHz) using a Hielscher (UIP500hd) sonicator (probe tip diameter of 22 mm) for 1 h. The solution was further left to dry and centrifuged (as described above) for the collection of the final product. The cobalt-functionalised MWCNTs (Co-f-MWCNTs) were stored in a desiccator at room temperature.

For the immobilisation of the enzyme on the electrode, the following procedure was followed: A total of 0.005 g of Co-f-MWCNTs were suspended in 500 µL of Nafion^{TM} perfluorinated resin solution (concentration 5 wt% in lower aliphatic alcohols and water, Aldrich). Then, 8 µL of the enzyme were mixed with 2 µL of Co-f-MWCNTs suspension. Finally, 2 µL of the resulting mix were left to dry on the GC electrode surface.

*Mt*LPMO9H was produced and purified to homogeneity as previously described [22]. After the purification the enzyme was divided in aliquots of 500 µL, and stored at −15 °C until use. Each aliquot was stored for a maximum time of 2 days at 4 °C after its first use. The concentration of the purified enzyme was about 5.5 mg/mL, identified by the method introduced by Lowry et al. [23].

3. Results

3.1. Cyclic Voltammetry

As a first step in the examination of the interaction of the immobilised *Mt*LPMO9H with the electrode surface via direct electron transfer, cyclic voltammetry was performed. *Mt*LPMO9H was immobilised on a Co-f-MWCNTs modified GC electrode and the potential was scanned between 0.6 and −0.1 V at a scan rate of 0.1 V s^{-1}. The voltammogram is presented in Figure 1 (black curve). Two rather faint peaks appear, one during the cathodic scan appearing at 0.252 V and one during the anodic scan at 0.281 V. Capacitance currents are evident in this voltammogram. The peak separation is estimated at 0.029 V, indicating a quasi-reversible reaction. The cyclic voltammetry of the Co-f-MWCNTs modified GC electrode in the absence of the enzyme is also present in Figure 1 (red curve). In this case, no redox peaks are detected, indicating that the redox peaks can indeed be attributed to the redox reaction of the enzyme and not any interference of the matrix.

In order to extract the kinetic constant k^0 and the transfer coefficient a from a cyclic voltammogram, assuming Butler–Volmer kinetics [24], one would take advantage of the correlation of the anodic and cathodic peak shift by varying the scan rate. However, these equations do not take the capacitance current and how it might affect the dependence of the peak shift into account. Moreover, at the higher scan rates which would be necessary in such a case, the capacitance would be more dominant, thus rendering the analysis even more difficult. One option would be to subtract the capacitance current from the total current in order to have only the Faradaic component and base the analysis on that alone. However, uncertainty rises for the precision of this approach in cases of non-ideal capacitance. The same uncertainty arises in case one would attempt to extract other magnitudes from the cyclic voltammogram, such as the surface coverage of the enzyme immobilised on the electrode surface.

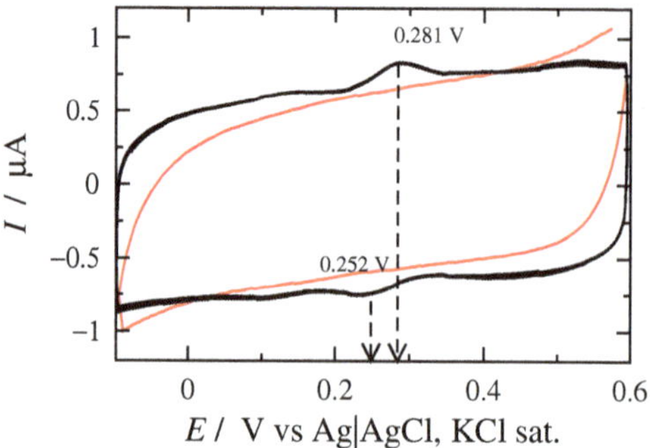

I / µA

E / V vs Ag|AgCl, KCl sat.

Figure 1. Experimental results: Cyclic voltammetry of *Mt*LPMO9H immobilised on a Co-f-MWCNTs modified GC electrode (black) and Co-f-MWCNTs modified GC electrode without enzyme (red) at 25 °C. Experimental conditions: acetate buffer 500 mM (pH 6) and $v = 0.1$ V s^{-1}.

Nonetheless, an interesting first approach to making an estimation of the value of the k^0 would be to solve the equations for the immobilised species numerically in the case of cyclic voltammetry for different standard electron transfer kinetic constants. In order to do so, the problem has to be formulated, by considering an electrochemical reaction having formal electrode potential $E^{0\prime}$, where both oxidised (E$_{Ox}$) and reduced (E$_{Red}$) form of the enzyme are immobilised on the electrode surface

$$\mathrm{E_{Ox}} + \mathrm{e} \underset{k_a}{\overset{k_c}{\rightleftharpoons}} \mathrm{E_{Red}} \tag{1}$$

where k_c the cathodic kinetic constant and k_a the anodic one. The units for both kinetic constants are 1/s. Let us denote as Γ^0 the initial surface concentration of E$_{Ox}$ species and assume a Butler–Volmer law for the rate constants

$$k_c = k^0 e^{-\frac{aF}{RT}(E-E^{0\prime})} \tag{2}$$

$$k_a = k^0 e^{\frac{(1-a)F}{RT}(E-E^{0\prime})} \tag{3}$$

Thus, the current density expression of the Faradaic current during the redox reaction of the immobilised species can be given from Equation (4) as a funtion of the surface coverage of the Ox species $\theta = \frac{\Gamma_{Ox}}{\Gamma^0}$, with Γ_{Ox} the surface coverage of the E$_{Ox}$ species

$$i_F = \Gamma^0 F k^0 \left[(1-\theta) e^{\frac{(1-a)F}{RT}(E-E^{0\prime})} - \theta e^{-\frac{aF}{RT}(E-E^{0\prime})} \right] \tag{4}$$

where F is the Faraday constant and the coverage is found by solving the following differential equation,

$$\frac{d\theta}{dt} = \frac{i_F}{F} \tag{5}$$

Assuming a cyclic voltammetry experiment where the potential starts from an initial value E_I and is scanned linearly towards more cathodic potentials against time at a certain scan rate v, and the scan is reversed at a time t_R, the applied potential can be expressed as in Equation (6)

$$E = E_I - v t_R + v|t - t_R| \tag{6}$$

In order to also include capacitance in the problem, we shall assume a contribution to the current, which can be expressed as follows

$$i_C = C_{dl} \frac{dE}{dt} \tag{7}$$

where C_{dl} is the specific capacitance of the double layer. The total current density, i, shall be expressed as a sum of the Faradaic and the capacitance contributions,

$$i(t) = i_F(t) + C_{dl}\left(v\frac{t - t_R}{|\,t - t_R\,|}\right) \tag{8}$$

By solving the problem numerically, the cyclic voltammograms are presented in Figure 2, for different k^0 values. The other parameteres used for this example are $\Gamma^0 = 5.1 \times 10^{-6}$ mol m^{-2}, $C_{dl} = 1$ F m^{-2}, $E^{0\prime} = 0$ V, $v = 0.1$ V s^{-1}. For the different k^0 values, the peak separation values are presented in Table 1. From these values, one would expect that based on the peak separation of the experimental results at 0.1 V s^{-1} at 0.029 V, a rough estimation of the k^0 value should be around 5 s^{-1}.

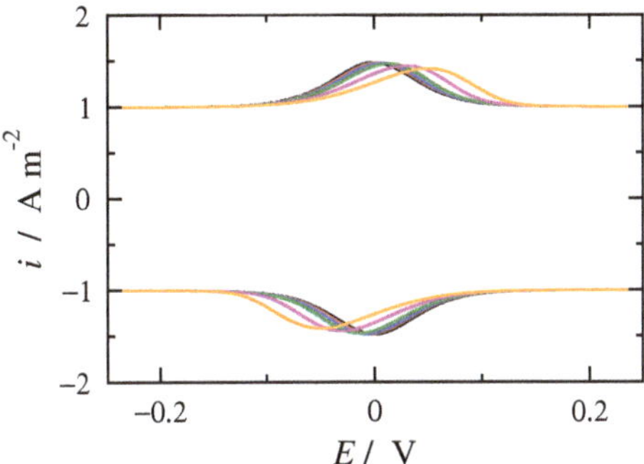

Figure 2. Computational results: Cyclic voltammetry for different k^0 values. Parameter values: $\Gamma^0 = 5.1 \times 10^{-6}$ mol m^{-2}, $C_{dl} = 1$ F m^{-2}, $E^{0\prime} = 0$ V, $v = 0.1$ V s^{-1} and $k^0 = 100$ (black), 20 (red), 10 (blue), 5 (green), 2 (orange) and 1 s^{-1} (orange).

Table 1. Peak separation values for an immobilised species assuming $v = 0.1$ V s^{-1} and ideal capacitance.

k^0/s^{-1}	ΔE/V
100	0.000
20	0.008
10	0.014
5	0.026
2	0.059
1	0.098

3.2. FT Alternating Current Voltammetry

Before attempting to extract the k^0, FTacV was performed on a Co-f-MWCNTs modified GC electrode with and without *Mt*LPMO9H. The cathodic 5th harmonic in acetate buffer 500 mM, pH 6 for $A_0 = 0.2$ V and $f = 5$ is presented in Figure 3. In the case of the immobilised enzyme (black curve) a set of intense peaks around the principle peak at 0.25 V. On the other hand, in the absence of the *Mt*LPMO9H, a faint peak of small amplitude appears at 0.2 V (red curve), leading to the conclusion that the high amplitude harmonic is indeed due to the enzyme and not the immobilization matrix.

Then, FTacV was performed so as to apply the methodology introduced [17] for the extraction of the k^0. At this point, the methodology is summarized:

1. A harmonic h is chosen, being free of capacitance currents. This harmonic is recorded at a fixed amplitude and decreasing frequencies;

2. The limiting value of the normalized peak height, $I_{p,h}^{(rev)}/f$, of the principal peak is determined. For this frequency, f, the reaction is at the reversible region;

3. If the transfer coefficient is to be determined, the shift in the potential of the principal peak is plotted as a function of the logarithm of the inverse frequency, $1/f$. Linear regression is performed in the region of high frequencies and the transfer coefficient is determined;

4. The principal peak height $I_{p,h}$ of a quasi-reversible harmonic is determined and the value $I_{p,h}/I_{p,h}^{(rev)}$ is calculated;

5. The kinetic constant is determined from the graph representing the dependance of $I_{p,h}/I_{p,h}^{(rev)}$ on $\log(k^0/f)$.

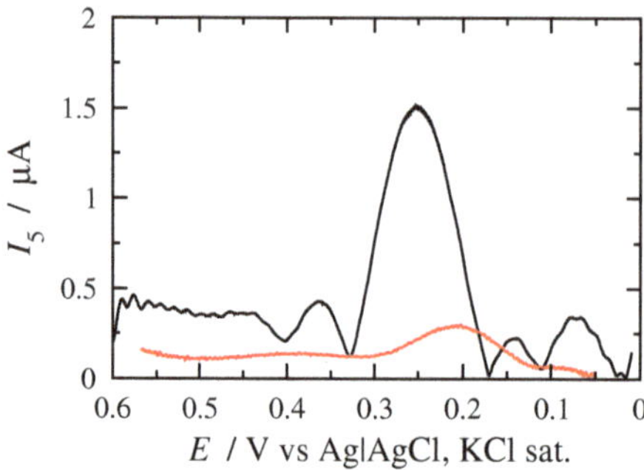

Figure 3. Experimental results: 5th harmonics of *Mt*LPMO9Himmobilised on a Co-f-MWCNTs modified GC electrode (black) and Co-f-MWCNTs modified GC electrode without the enzyme (red) at 25 °C. Experimental conditions: acetate buffer 500 mM, pH 6, $A_0 = 0.20$ V and $f = 5$.

Therefore, as a first step, FTacV was performed for two amplitudes, 0.2 and 0.25 V, at frequencies of 5, 3, 1, 0.5, 0.2 and 0.1 Hz. The scan rate was adjusted at each case accordingly in order to fulfill the slow scan rate approximation and render the experiment independent of this factor [16,19]. The cathodic scan of the 4th and the 5th harmonics are presented in Figures 4 and 5, respectively, for $A_0 = 0.2$ and 0.25 V. Looking at both the amplitudes employed, there is no intense deformation in the harmonics that would suggest a transfer coefficient that would deviate beyond the range of 0.4–0.6; thus, regarding the dominant peak [17], no significant errors would be inserted in the preliminary analysis in case an $a = 0.5$ is assumed. Moreover, both 4th and 5th harmonic seem free of capacitance contribution.

As a second step, the trend of the main observable, i.e., the peak height of the principle peaks, is to be observed. In both cases, typical 4th and 5th harmonics for quasi-reversible reactions of immobilised species are presented for the higher applied frequencies (Figures 4a,c and 5a,c). Apparently, as the frequency is decreased, the amplitude of the harmonics decreases, as predicted theoretically. The trend of the normalized peak height of the principle peaks, $I_{p,h}/f$, on frequency is presented in Figures 4b,d and 5b,d. By observing these figures, it is evident that a saturation in the peak height of the dominant peak is observed for a frequency of 0.1 Hz. This indicates that reversibility is reached by that point, and the height of the dominant peak of the reversible harmonic, $I_{p,h}^{(rev)}$ has been determined.

As a next step, a quasi-reversible harmonic is chosen and the ratio $I_{p,h}/I_{p,h}^{(rev)}$ is calculated, where $I_{p,h}$ is the normalised peak current for a quasi-reversible harmonic and $I_{p,h}^{(rev)}$ the normalised peak current for the reversible harmonic at a certain A_0 and a. For all the combinations of amplitudes and harmonics, the chosen frequency for the quasi-reversible harmonic is that of 5 Hz, while for the peak of the reversible harmonic, it is 0.1 Hz. Let us now use the example of $I_{p,4}/I_{p,4}^{(rev)}$ and $I_{p,5}/I_{p,5}^{(rev)}$ for 0.25 V. The respective estimated values are 0.57 and 0.6, respectively.

As a final step, these values shall be used for the estimation of the k^0 from the nomograms of $I_{p,h}/I_{p,h}^{(rev)}$ against the $\log(k^0 f^{-1})$. These nomograms are calculated for $a = 0.5$ (Figure 6a for the 4th harmonic and Figure 6b for the 5th). The same procedure is done for the amplitude of $A_0 = 0.2$ V. By taking an average of the extracted values, the k^0 is estimated at 4.6 ± 2.6 s^{-1}.

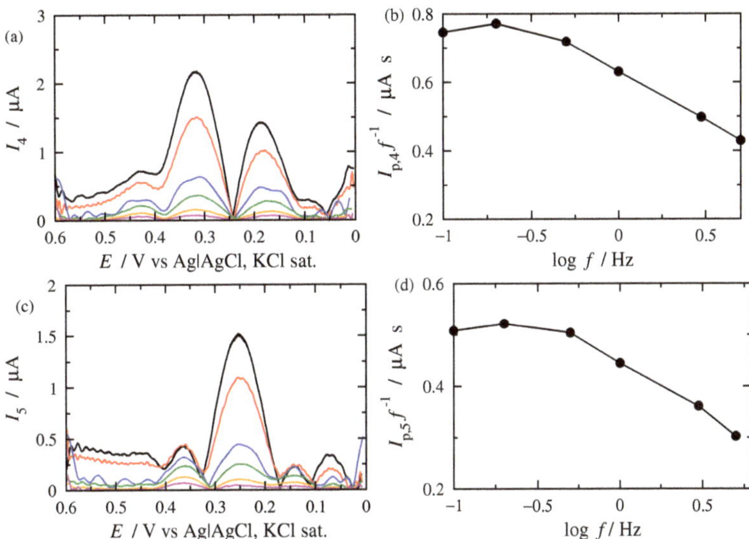

Figure 4. Experimental results: (**a**) 4th harmonics, (**b**) normalised peak height of the 4th harmonic, (**c**) 5th harmonics and (**d**) normalised peak height of the 5th harmonic, of *Mt*LPMO9H immobilised on a Co-f-MWCNTs modified GC electrode at 25 °C. Experimental conditions: acetate buffer 500 mM, pH 6, $A_0 = 0.2$ V, and $f = 5$ (black), 3 (red), 1 (blue), 0.5 (green), 0.2 (orange) and 0.1 (magenta) Hz.

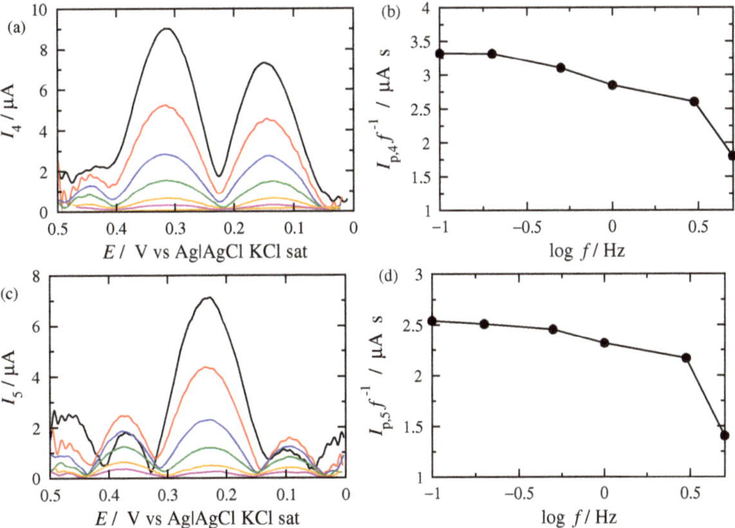

Figure 5. Experimental results: (**a**) 4th harmonics, (**b**) normalised peak height of the 4th harmonic, (**c**) 5th harmonics and (**d**) normalised peak height of the 5th harmonic, of *Mt*LPMO9H immobilised on a Co-f-MWCNTs modified GC electrode at 25 °C. Experimental conditions: acetate buffer 500 mM, pH 6, $A_0 = 0.25$ V, and $f = 5$ (black), 3 (red), 1 (blue), 0.5 (green), 0.2 (orange) and 0.1 (magenta) Hz.

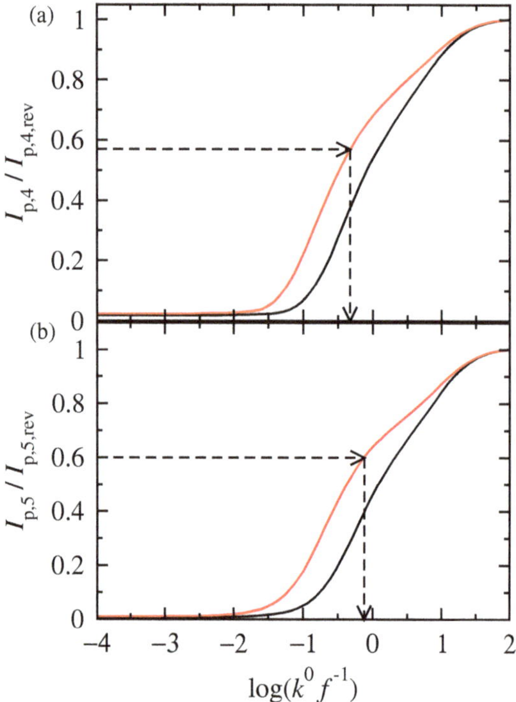

Figure 6. Computational results: Dependence of the normalised principal peak height for the (**a**) 4th and (**b**) 5th harmonic of an immobilised species. Parameters: $a = 0.5$ and $A_0 = 0.2$ (black) and 0.25 V (red).

In order to validate the value of k^0, found experimentally, FTacV experiments were simulated numerically for k^0 equal to 4.6 s^{-1} and a equal to 0.57. In order to do so, Equation (5) was solved numerically where the potential ramp is perturbed by a sinusoidal of amplitude A_0 and frequency f,

$$E = E_I - vt_R + v|t - t_R| + A_0 \sin 2\pi f t \qquad (9)$$

and the harmonics where calculated from the current obtained by multiplying the current density, Equation (4) with the electrode surface S. Apparently, both the surface area S and the surface concentration of the oxidized form of the enzyme Γ^0 are unknown and are expected to vary between different immobilisation experiments. Therefore, the factor $\Gamma^0 S$ was chosen by trial-and-error. The simulation results (red curves) together with the corresponding experimental plots (black curves) for $A_0 = 0.2$, $\Gamma^0 S = 3 \times 10^{-13}$ mol and $A_0 = 0.25$ V, $\Gamma^0 S = 1.2 \times 10^{-12}$ mol and $f = 1$ Hz, are presented in Figure 7a,b, respectively. A rather good agreement is observed, implying a satisfactory estimation of k^0.

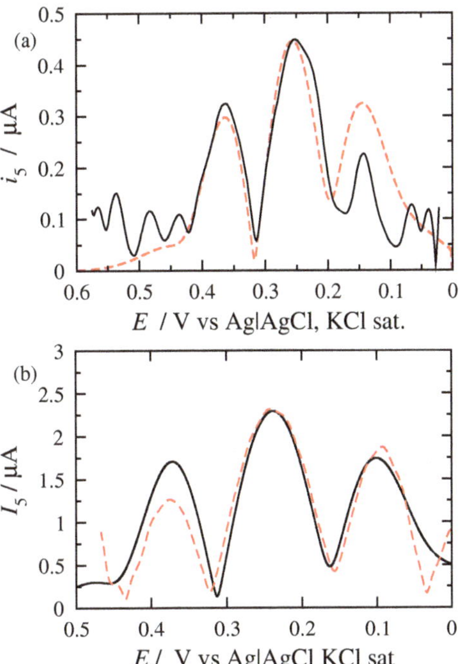

Figure 7. Experimental (black) versus computational (red dashed line) of the 5th harmonic of *Mt*LPMO9H immobilised on a Co-f-MWCNTs modified GC electrode at 25 °C. Experimental conditions: acetate buffer 500 mM, pH 6, $f = 1$ Hz and (**a**) $A_0 = 0.2$ and (**b**) 0.25 V. Computational parameters: a=0.57, $f = 1$ Hz, (**a**) $A_0 = 0.2$ V, $\Gamma^0 S = 3 \times 10^{-13}$ mol and (**b**) $A_0 = 0.25$ V, $\Gamma^0 S = 1.2 \times 10^{-12}$ mol.

4. Discussion

*Mt*LPMO9H was immobilised on an electrode surface, ensuring that direct electron transfer is indeed possible for this enzyme with the immobilisation method used with an indication that, for the conventional scan, rated when using cyclic voltammetry, the kinetics were quasi-reversible. The signal was governed by capacitance, a common feature for immobilised enzymes on electrode surfaces. Despite this, still, an initial estimation of the standard electron transfer rate constant was intented through the computational reproduction of cyclic voltammograms. From the peak separation, it was predicted that the k^0 would be around 5 s^{-1}, assuming an *a* equal to 0.5.

Then FTacV was performed for two different amplitudes as various frequencies in order to assure that both quasi-reversibility and reversibility are achieved in order to apply the methodology to extract the k^0 of immobilised species using FTacV. The extracted value was at 4.6 ± 2.6 s^{-1} being in good accordance with what was expected from cyclic voltammetry.

In order to validate the k^0 value obtained experimentally, FTacV experiments were simulated numerically and a rather good agreement was found for transfer coefficient *a* equal to 0.57. In order to approximate the peak heights of the voltammograms, different values of $\Gamma^0 S$ were used for the two different amplitudes. At this point, the validity of the $\Gamma^0 S$ used shall be ascertained. Since the enzyme concentration is 5.5 mg mL^{-1} and 8 µL were utilised, and they were dilluted in 10 µL to a final concentration of 4.4 mg mL^{-1}. From this, 2 µL were used, and, taking into consideration that the molecular weight of the enzyme is 65 kDa [22], the maximum moles that can be 1.35×10^{-10}. Considering that not all of the enzymes would be bound to the matrix and from what was bound, not all would be oriented in such a way so as to interact with the enzyme, the values used for the simulation seem to be valid. Nevertheless, it has to be pointed out that the proposed experimental methodology for obtaining kinetic parameters does not require knowledge of the surface concentration of the immobilised species or the the value of the surface area.

In general, it is common practice to compare computational with experimental results in order to ascertain the validity of the parameters used for the computational ones. The computational fits tend to reproduce most of the features of the voltammogram, just as in this case, where the basic features of the harmonics presented are fitted rather well [25,26]. Regarding the precision of the comparison of the computational harmonics with the experimental ones, it should also be taken into account that the above analysis is valid only if the assumptions of the model are closely met, where the immobilised species must behave as a Langmuirian monolayer and a dispersion of the electrochemical transfer rates should be absent. Otherwise, where kinetic or thermodynamic dispersion is present, more elaborate models have to be considered [27–29]. The influence of kinetic and thermodynamic dispersion might affect the waveform of the voltammograms [30], and other theories such as that of Marcus–Hush can applied, instead of the Butler–Volmer law [31]. Yet, in this case, the Butler–Volmer law seems to be rather sufficient for this case.

To conclude, with the k^0 value that was estimated, it seems that, under the right conditions, it is feasible to render the system reversible and take advantage of it for bioelectrocatalytic purposes. Thus, it could be proven to be rather resourceful in such applications.

Author Contributions: Conceptualization, D.Z. and A.K. (Antonis Karantonis); methodology, D.Z.; software, A.K. (Antonis Karantonis); validation, D.Z. and A.K. (Antonis Karantonis); formal analysis, R.X., D.Z.; investigation, D.Z.; resources, A.K. (Antonis Karantonis), E.T.; data curation, D.Z.; writing—original draft preparation, D.Z., A.K. (Antonis Karantonis); writing—review and editing, A.K. (Antonis Karantonis), A.K. (Anthi Karnaouri), E.T.; supervision, A.K. (Antonis Karantonis), E.T.; project administration, A.K. (Antonis Karantonis); funding acquisition, A.K. (Antonis Karantonis). All authors have read and agreed to the published version of the manuscript.

Funding: The research work was supported by the Hellenic Foundation for Research and Innovation (H.F.R.I.) under the "First Call for H.F.R.I. Research Projects to support Faculty members and Researchers and the procurement of high-cost research equipment grant" (Project Number: HFRI-FM17-3090).

Institutional Review Board Statement: Not applicable.

Informed Consent Statement: Not applicable.

Data Availability Statement: Data sharing not applicable.

Acknowledgments: Chr. Argirusis and P. Pandis are aknowledged for providing the Co-f-MWCNTs.

Conflicts of Interest: The authors declare no conflict of interest.

References

1. Eibinger, M.; Ganner, T.; Bubner, P.; Rošker, S.; Kracher, D.; Haltrich, D.; Ludwig, R.; Plank, H.; Nidetzky, B. Cellulose surface degradation by a lytic polysaccharide monooxygenase and its effect on cellulase hydrolytic efficiency. *J. Biol. Chem.* **2014**, *289*, 35929–35938. [CrossRef] [PubMed]
2. Zhang, R. Functional characterization of cellulose-degrading AA9 lytic polysaccharide monooxygenases and their potential exploitation. *Appl. Microbiol. Biotechnol.* **2020**, *104*, 3229–3243. [CrossRef]
3. Moreau, C.; Tapin-Lingua, S.; Grisel, S.; Gimbert, I.; Le Gall, S.; Meyer, V.; Petit-Conil, M.; Berrin, J.-G.; Cathala, B.; Villares, A. Lytic polysaccharide monooxygenases (LPMOs) facilitate cellulose nanofibrils production. *Biotechnol. Biofuels* **2019**, *12*, 1–13. [CrossRef]
4. Forsberg, Z.; Stepnov, A.A.; Kruge Nærdal, G.; Klinkenberg, G.; Eijsink, V. G.H. Engineering lytic polysaccharide monooxygenases (LPMOs). *Method Enzymol.* **2020**, *644*, 1–34.
5. Hu, J.; Arantes, V.; Pribowo, A; Saddler, J. N. The synergistic action of accessory enzymes enhances the hydrolytic potential of a "cellulase mixture" but is highly substrate specific. *Biotechnol. Biofuels* **2013**, *6*, 112. [CrossRef] [PubMed]
6. Zerva, A.; Pentari, C.; Termentzi, A.; America, A.H.P.; Zouraris, D.; Bhattacharya, S.K.; Karantonis, A.; Zervakis, G.I.; Topakas, E. Discovery of two novel laccase-like multicopper oxidases from *Pleurotus citrinopileatus* and their application in phenolic oligomer synthesis. *Biotechnol. Biofuels* **2021**, *14*, 83. [CrossRef]

7. Frandsen, K.E.H.; Lo Leggio, L. Lytic polysaccharide monooxygenases: A crystallographer's view on a new class of biomass-degrading enzymes. *IUCrJ* **2016**, *3*, 448–467. [CrossRef]

8. Bissaro, B.; Røhr, Å.K.; Müller, G.; Chylenski, P.; Skaugen, M.; Forsberg, Z.; Horn, S.J.; Vaaje-Kolstad, G.; Eijsink, V.G.H. Oxidative cleavage of polysaccharides by monocopper enzymes depends on H_2O_2. *Nat. Chem. Biol.* **2017**, *13*, 1123. [CrossRef]

9. Vaaje-Kolstad, G.; Westereng, B.; Horn, S.J.; Liu, Z.; Zhai, H.; Sørlie, M.; Eijsink, V.G.H. An oxidative enzyme boosting the enzymatic conversion of recalcitrant polysaccharides. *Science* **2010**, *330*, 219–222. [CrossRef]

10. Zerva, A.; Pentari, C.; Grisel, S; Berrin, J.-G.; Topakas, E. A new synergistic relationship between xylan-active LPMO and xylobiohydrolase to tackle recalcitrant xylan. *Biotechnol. Biofuels* **2020**, *13*, 142. [CrossRef] [PubMed]

11. Cannella, D.; Chia-wen, C.H.; Felby, C.; Jørgensen, H. Production and effect of aldonic acids during enzymatic hydrolysis of lignocellulose at high dry matter content. *Biotechnol. Biofuels* **2012**, *5*, 1–10. [CrossRef] [PubMed]

12. Frommhagen, M.; Westphal, A.H.; Van Berkel, W.J.H.; Kabel, M.A. Distinct substrate specificities and electron-donating systems of fungal lytic polysaccharide monooxygenases. *Front. Microbiol.* **2018**, *9*, 1080. [CrossRef] [PubMed]

13. Span, E.A.; Marletta, M.A. The framework of polysaccharide monooxygenase structure and chemistry. *Curr. Opin. Struct. Biol.* **2015**, *35*, 93–99. [CrossRef]

14. Chen, H.; Dong, F.; Minteer, S.D. The progress and outlook of bioelectrocatalysis for the production of chemicals, fuels and materials. *Nat. Catal.* **2020**, *3*, 225–244. [CrossRef]

15. Masa, J.; Schuhmann, W. Electrocatalysis and bioelectrocatalysis–Distinction without a difference. *Nano Energy* **2016**, *29*, 466–475. [CrossRef]

16. Zouraris, D.; Karantonis, A. Large amplitude ac voltammetry: Chief observables for a reversible reaction of free electroactive species. *J. Electroanal. Chem.* **2019**, *847*, 113245. [CrossRef]

17. Zouraris, D.; Karantonis, A. Determination of kinetic and thermodynamic parameters from large amplitude Fourier transform ac voltammetry of immobilised electroactive species. *J. Electroanal. Chem.* **2020**, *876*, 114729. [CrossRef]

18. Engblom, S.O.; Myland, J.C.; Oldham, K.B.; Taylor, A.L. Large amplitude ac voltammetry—A comparison between theory and experiment. *Electroanalysis* **2001**, *13*, 626–630. [CrossRef]

19. Bell, C.G.; Anastassiou, C.A.; O'Hare, D.; Parker, K.H.; Siggers, J.H. Large-amplitude ac voltammetry: Theory for reversible redox reactions in the "slow scan limit approximation". *EElectrochim. Acta* **2011**, *56*, 6131–6141. [CrossRef]

20. Zouraris, D.; Dimarogona, M.; Karnaouri, A.; Topakas, E.; Karantonis, A. Direct electron transfer of lytic polysaccharide monooxygenases (LPMOs) and determination of their formal potentials by large amplitude Fourier transform alternating current cyclic voltammetry. *Bioelectrochemistry* **2018**, *124*, 149–155. [CrossRef]

21. Zouraris, D.; Kiafi, S.; Zerva, A.; Topakas, E.; Karantonis, A. FTacV study of electroactive immobilized enzyme/free substrate reactions: Enzymatic catalysis of epinephrine by a multicopper oxidase from *Thermothelomyces thermophila*. *Bioelectrochemistry* **2020**, *134*, 107538. [CrossRef] [PubMed]

22. Karnaouri, A.; Muraleedharan, M.N.; Dimarogona, M.; Topakas, E.; Rova, U.; Sandgren, M.; Christakopoulos, P. Recombinant expression of thermostable processive *Mt*EG5 endoglucanase and its synergism with *Mt*LPMO from *Myceliophthora thermophila* during the hydrolysis of lignocellulosic substrates. *Biotechnol. Biofuels* **2017**, *10*, 1–17. [CrossRef] [PubMed]

23. Lowry, O.H.; Rosebrough, N.J.; Farr, A.L.; Randall, R.J. Protein measurement with the Folin phenol reagent. *J. Biol. Chem.* **1951**, *193*, 265–275. [CrossRef]

24. Savéant, J.-M.; Costentin, C. *Elements of Molecular and Biomolecular Electrochemistry: An Electrochemical Approach to Electron Transfer Chemistry*, 2nd ed.; John Wiley & Sons Inc.: Hoboken, NJ, USA, 2019.

25. Zhang, Y.; Simonov, A.N.; Zhang, J.; Bond, A.M. Fourier transformed alternating current voltammetry in electromaterials research: Direct visualisation of important underlying electron transfer processes. *Curr. Opin. Electrochem.* **2018**, *10*, 72–81 [CrossRef]

26. Song, P.; Ma, H.; Meng, L.; Wang, Y.; Nguyen, H.V.; Lawrence, N.S.; Fisher, A.C. Fourier transform large amplitude alternating current voltammetry investigations of the split wave phenomenon in electrocatalytic mechanisms. *Phys. Chem. Chem. Phys.* **2017**, *19*, 24304–24315. [CrossRef] [PubMed]

27. Léger, C.; Jones, A.K.; Albracht, S.P.J. Armstrong, F.A. Effect of a dispersion of interfacial electron transfer rates on steady state catalytic electron transport in [NiFe]-hydrogenase and other enzymes. *J. Phys. Chem. B* **2002**, *106*, 13058–13063. [CrossRef]

28. Robinson, M.; Ounnunkad, K.; Zhang, J.; Gavaghan, D.; Bond, A.M. Models and their limitations in the voltammetric parameterization of the six-electron surface-confined reduction of $[PMo_{12}O_{40}]^{-3}$ at glassy carbon and boron-doped diamond electrodes. *ChemElectrochem* **2019**, *6*, 5499–5510. [CrossRef]

29. Rahman, M.A.; Guo, S.-X.; Laurans, M.; Izzet, G.; Proust, A.; Bond, A.M.; Zhang, J. Thermodynamics, electrode kinetics, and mechanistic nuances associated with the voltammetric reduction of dissolved $[n\text{-}Bu_4N]_4[PW_{11}O_{39}\{Sn(C_6H_4)C\equiv C(C_6H_4)(N_3C_4H_{10})\}]$ and a surface-confined diazonium derivative. *ACS Appl. Energy Mater.* **2020**, *3*, 13991–14006. [CrossRef]

30. Morris, G.P.; Baker, R.E.; Gillow, K.; Davis, J.J.; Gavaghan, D.J.; Bond, A.M. Theoretical analysis of the relative significance of thermodynamic and kinetic dispersion in the dc and ac voltammetry of surface-confined molecules. *Langmuir* **2015**, *31*, 4996–5004. [CrossRef]

31. Stevenson, G.P.; Baker, R.E.; Kennedy, G.F.; Bond, A.M.; Gavaghan, D.J.; Gillow, K. Access to enhanced differences in Marcus-Hush and Butler-Volmer electron transfer theories by systematic analysis of higher order AC harmonics. *Phys. Chem. Chem. Phys.* **2013**, *15*, 2210–2221. [CrossRef]

Article

Numerical Investigation of Latent Thermal Storage in a Compact Heat Exchanger Using Mini-Channels

Amir Abdi *, Justin Ningwei Chiu and Viktoria Martin

KTH Royal Institute of Technology, 100 44 Stockholm, Sweden; justin.chiu@energy.kth.se (J.N.C.);
viktoria.martin@energy.kth.se (V.M.)
* Correspondence: aabdi@kth.se

Abstract: This paper aims to numerically investigate the thermal enhancement of a latent thermal energy storage component with mini-channels as air passages. The investigated channels in two sizes of internal air passages (channel-1 with d_h = 1.6 mm and channel-2 with d_h = 2.3 mm) are oriented vertically in a cuboid of $0.15 \times 0.15 \times 0.1$ m^3 with RT22 as the PCM located in the shell. The phase change is simulated with a fixed inlet temperature of air, using ANSYS Fluent 19.5, with a varying number of channels and a ranging air flow rate entering the component. The results show that the phase change power of the LTES improves with by increasing the number of channels at the cost of a decrease in the storage capacity. Given a constant air flow rate, the increase in the heat transfer surface area of the increased number of channels dominates the heat transfer coefficient, thus increasing the mean heat transfer rate (UA). A comparison of the channels shows that the thermal performance depends largely on the area to volume ratio of the channels. The channel type two (channel-2) with a slightly higher area to volume ratio has a slightly higher charging/discharging power, as compared to channel type one (channel-1), at a similar PCM packing factor. Adding fins to channel-2, doubling the surface area, improves the mean UA values by 15–31% for the studied cases. The variation in the total air flow rate from 7 to 24 L/s is found to have a considerable influence, reducing the melting time by 41–53% and increasing the mean UA values within melting by 19–52% for a packing factor range of 77.4–86.8%. With the increase in the air flow rate, channel type two is found to have considerably lower pressure drops than channel type one, which can be attributed to its higher internal hydraulic diameter, making it superior in terms of achieving a relatively similar charging/discharging power in exchange for significantly lower fan power. Such designs can further be optimized in terms of pressure drop in future work, which should also include an experimental evaluation.

Keywords: PCM; mini-channels; air; melting; solidification

Citation: Abdi, A.; Chiu, J.N.;
Martin, V. Numerical Investigation of
Latent Thermal Storage in a Compact
Heat Exchanger Using Mini-
Channels. *Appl. Sci.* **2021**, *11*, 5985.
https://doi.org/10.3390/
app11135985

Academic Editor: Ioannis Kartsonakis

Received: 28 May 2021
Accepted: 24 June 2021
Published: 27 June 2021

Publisher's Note: MDPI stays neutral
with regard to jurisdictional claims in
published maps and institutional affiliations.

1. Introduction

Integration of thermal energy storage in ventilation systems can be advantageous to decrease system-associated costs, including capital and operational, and contribute to peak power savings [1,2]. Latent Thermal Energy Storage (LTES) is a favorable solution due to the large storage capacity of Phase Change Materials (PCM) and its low volume. The major obstacle associated with the LTES is the low thermal conductivity of the Phase Change Materials [3]. This results in a low charging/discharging power, requiring enhancement in heat transfer mechanisms [4]. The thermal performance of such a system is intensified even further in cases where air is used as the heat transfer fluid (HTF) [5].

The thermal performance of LTES systems working with air was addressed in some research [6–10]. Dolado et al. [6], working on a real scale air–PCM heat exchanger, pointed out that the bottleneck of the heat transfer, particularly at low air flow rates, could be due to the high resistance on the air side. The analysis showed that the system performance was improved by increasing the flow rate of air or decreasing the air channel thickness; however, at the cost of having a higher fan energy consumption.

Vakilaltojjar and Saman [7] analyzed a plate encapsulating PCM system working with air and reported that the thermal performance could be improved by decreasing the dimensions of the air channels and the PCM slabs.

Marin et al. [8] investigated an LTES system, enhanced with graphite, and concluded that in cases with the PCM–graphite composite, as the storage medium, the thermal performance of the LTES was controlled by the resistance on the air side, due to the reduced resistance of the storage medium. On the other hand, in cases where pure PCM was employed, the resistances on the PCM and the airside were competitive.

Herbinger et al. [10] performed a parametric study of an air–PCM heat exchanger for ventilation systems, considering square channels of 2×2 cm^2, 4×4 cm^2 and 6×6 cm^2. They found that as the channel dimension decreased, the charging/discharging power increased, and the best thermal performance was achieved for the smallest channels. However, in the study, the influence of reducing the hydraulic diameter of the channels and subsequently the increase in the heat transfer coefficient of air and its effect on the thermal performance was lacking.

The hydraulic diameter (d_h) of the flow passages can be reduced significantly by using mini- and micro-channels. Mini/micro-channels proved to be significantly beneficial in various industries and applications such as the refrigeration/heat pump industry and electronics cooling [11–14]. Kandlikar et al. [11] classified the passages based on the dimensions of the hydraulic diameter (d_h) to conventional channels for $d_h > 3$ mm, mini-channels for 0.2 mm $< d_h < 3$ mm and micro-channels for 0.01 mm $< d_h < 0.2$ mm. Kandlikar et al. [11] showed that the fluid heat transfer coefficients for circular passages are inversely proportional to the hydraulic diameter, given a fixed flow rate. However, the increase in heat transfer coefficient is achieved at a considerable cost of an increase in the pressure drop. Besides the hydraulic diameter, with rectangular channels, the aspect ratio of the channels as a fraction of the width to the height can influence the heat transfer coefficient. Wibulawas [15], in a study on laminar flow in heat transfer in non-circular ducts, showed that, given a constant Graetz number ($Gz = \frac{d_h}{l} RePr$), the Nusselt number either under constant heat flux or constant wall temperature conditions increases with the increase in the aspect ratios of the channels.

Rectangular mini-channels with low hydraulic diameters and high area-to-volume ratios have the potential to enhance the heat transfer on both the air and PCM sides with having relatively high PCM packing factors (PF). This is particularly beneficial in LTES applications related to air as the HTF, where there are needs for enhancement due to low thermal conductivities of air and PCMs.

Mini/micro-channels can be fabricated conventionally using extrusion or Additive Manufacturing (AM) processes [16–20]. However, an extrusion process is associated with certain limitations and restrictions in producing complex features. Additive Manufacturing (AM), known as 3D printing, is an alternative new fabrication method capable of manufacturing complex geometric features, including the embedding of internal/external thermal enhancers and bypassing many limitations associated with conventional methods. However, 3D printing is also associated with certain disadvantages such as higher surface roughness, limited options in printing materials, and higher cost per part in mass productions, as compared to the conventional methods [21].

This paper numerically investigates the novel idea of incorporating mini-channel profiles into LTES components with dry air as the HTF. Three-dimensional simulations of transient phase change were performed using ANSYS Fluent 19.5. An enthalpy–porosity method is used in this commercial software for solid–liquid phase change simulations. In this method, the mushy zone at which the phase change is happening is dealt with as a porous structure, and the liquid flow in this zone is treated as a liquid flow through a porous structure [22,23].

Two types of mini-channels were studied. The first channel is an existing extruded aluminum profile and the second channel, with and without external fins, is a conceptual design capable of being manufactured via 3D printing. The mini-channels have relatively similar volumes but different internal hydraulic diameters ($d_{h\,channel-1}$ = 1.6 and $d_{h\,channel-1}$ = 2.3) and aspect ratios ($AR_{channel-1}$ = 0.86 and $AR_{channel-2}$ = 20). The compact LTESs are modeled three-dimensionally for a laminar air flow regime. The thermal performance of the LTES in charging/discharging with varying number of channels and ranging flow rates was analyzed and evaluated.

2. Materials and Methods

The numerical model to simulate melting and solidification processes of RT22 in an air–PCM heat exchanger using mini-channels is presented and described below.

2.1. Physical Model

Simulations of melting and solidification processes of RT22 as PCM were performed with two types of mini-channels. The air–PCM heat exchanger, as shown in Figure 1, is a cuboid of $0.15 \times 0.15 \times 0.1$ m^3 comprised of a duct shell and the mini-channels. The PCM is located in the shell, and air, as the HTF, flows through mini-channels. To reduce the computational effort of modeling the phase change process, a symmetric domain of one channel was simulated.

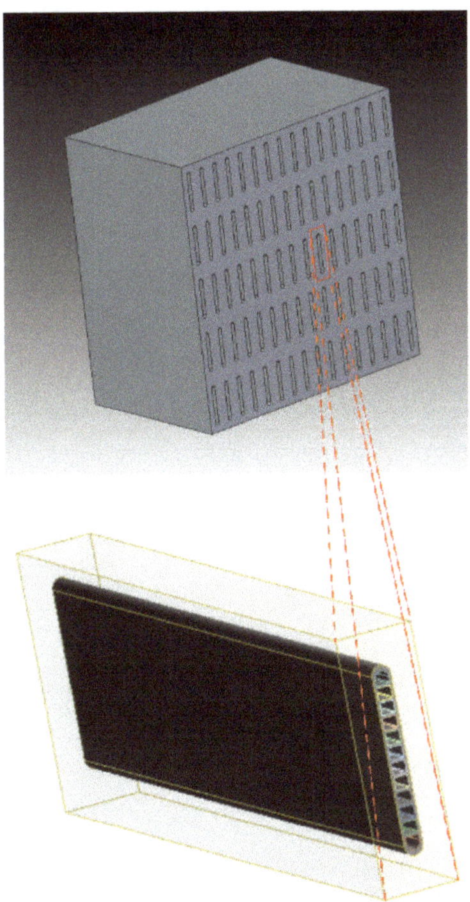

Figure 1. The physical model and the simulation domain.

Two types of channels, with specified dimensions shown in Figure 2, were considered in the study. The mini-channel type one (channel-1) is an existing product of Hydro Aluminum, manufactured through an extrusion and studied for the first time in an LTES component. Channel-1 has hydraulic diameters of 1.63 mm and 1.59 mm for the middle and the side passages, respectively. The mini-channel type two (channel-2) is a conceptual design with the purpose of improving the thermal performance of channel-1. This channel could be potentially manufactured via metal 3D printing as a fast and quick production method for innovative prototypes, using $AlSi_{10}Mg$ alloy as the material, with and without external fins. Channel-2 has a slightly higher area-to-volume ratio ($\frac{A_s}{Vol}$) than channel-2, providing a higher heat transfer surface given a similar PCM packing factor. Channel-2 was proposed with a hydraulic diameter and aspect ratio higher than Channel 1 to achieve a potentially higher heat transfer coefficient of air and a lower pressure drop. It has a hydraulic diameter of 2.27 mm and an aspect ratio of 17.5. The geometrical characteristics of the channels with their material properties are summarized in Table 1.

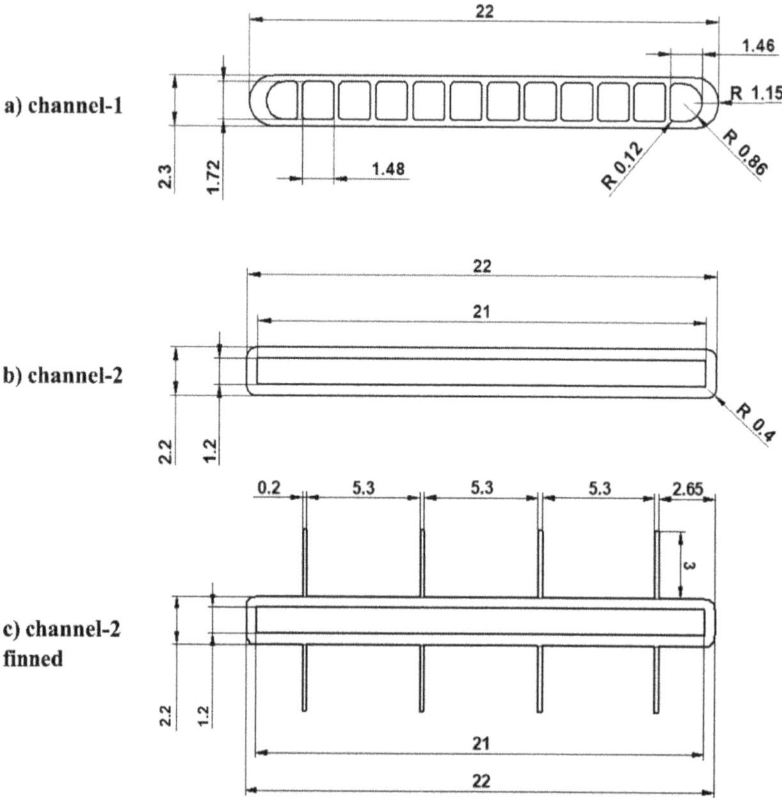

Figure 2. Dimensions of the simulated configurations: (**a**) channel-1, (**b**) channel-2 and (**c**) channel-2 finned.

Table 1. Geometrical characteristics and material properties of the channels.

	Channel-1 (Extrusion-Al3003)	Channel-2 (DMLS-AlSi$_{10}$Mg)
Area to volume ratio $\left[\frac{A_s}{Vol} \left(\frac{m^2}{m^3} \right) \right]$	943	987
Hydraulic diameter [d_h (mm)]	1.63	2.27
Density ($\frac{kg}{m^3}$)	2730	2620
Specific heat capacity ($\frac{kJ}{kg \cdot K}$)	892	846
Thermal conductivity ($\frac{W}{m \cdot K}$)	162	111

In the cuboid, the mini-channels with varying quantities are oriented vertically. In the case of channel-1, due to limitations in the production procedures of the cuboid, including the restrictions associated with brazing and welding processes, the channels need to maintain a minimum distance of 5 mm. In the case of channel-2, due to the inherited characteristics of the 3D-printing method, this distance could be lowered down to 1–2 mm. Thus, the number of channels per the height of the cuboid was set to five and six for channel-1 and channel-2, respectively. In the case of externally finned channel-2, four fins with a length of 3 mm and thickness of 0.2 mm on each side of the channel-2 were considered, almost doubling the external surface area. The number of channels in width of the cuboid is varied from 12 to 20 with packing factor (PF), defined as a ratio of the PCM volume to the total volume of the cuboid, range of 77.4–86.8%. The details of simulated configurations and the corresponding PFs are summarized in Table 2.

Table 2. Investigated configurations.

	Number of Channels	PF (%)
Channel-1	5 × 12	86.8
	5 × 16	82.4
	5 × 20	78
Channel-2	6 × 12	84.6
	6 × 14	82
	6 × 16	79.4
Channel-2-finned	6 × 12	83
	6 × 14	80.2
	6 × 16	77.4

2.2. PCM Thermos-Physical Properties

Material properties of the investigated PCM and RT22 are missing in the literature. RT22 is an organic commercial PCM provided by RUBITHERM undergoing the solid–liquid phase change around 22 °C. The thermo-physical properties, including density in the liquid phase, enthalpy and isobaric specific heat capacity, thermal conductivity and viscosity, were measured experimentally. The liquid density was measured using two pyknometers of 50 and 100 mL, enthalpy and specific heat via a μDSC evo 7 from the Setaram instrument, thermal conductivity via a Hot Disk TPS-2500 and viscosity via a rotational Brookfield viscometer. The linear correlation of the measured stress with the strain rate in the liquid phase, obtained in the viscosity measurement procedure, shows that RT22 has a Newtonian behavior. The measured properties are shown in Figure 3 as functions of temperature and summarized in Table 3. The measurement methodology via the aforementioned instruments used to measure the thermo-physical properties of RT22 in this paper was explained in detail by Abdi et al. [24] for two other organic PCMs.

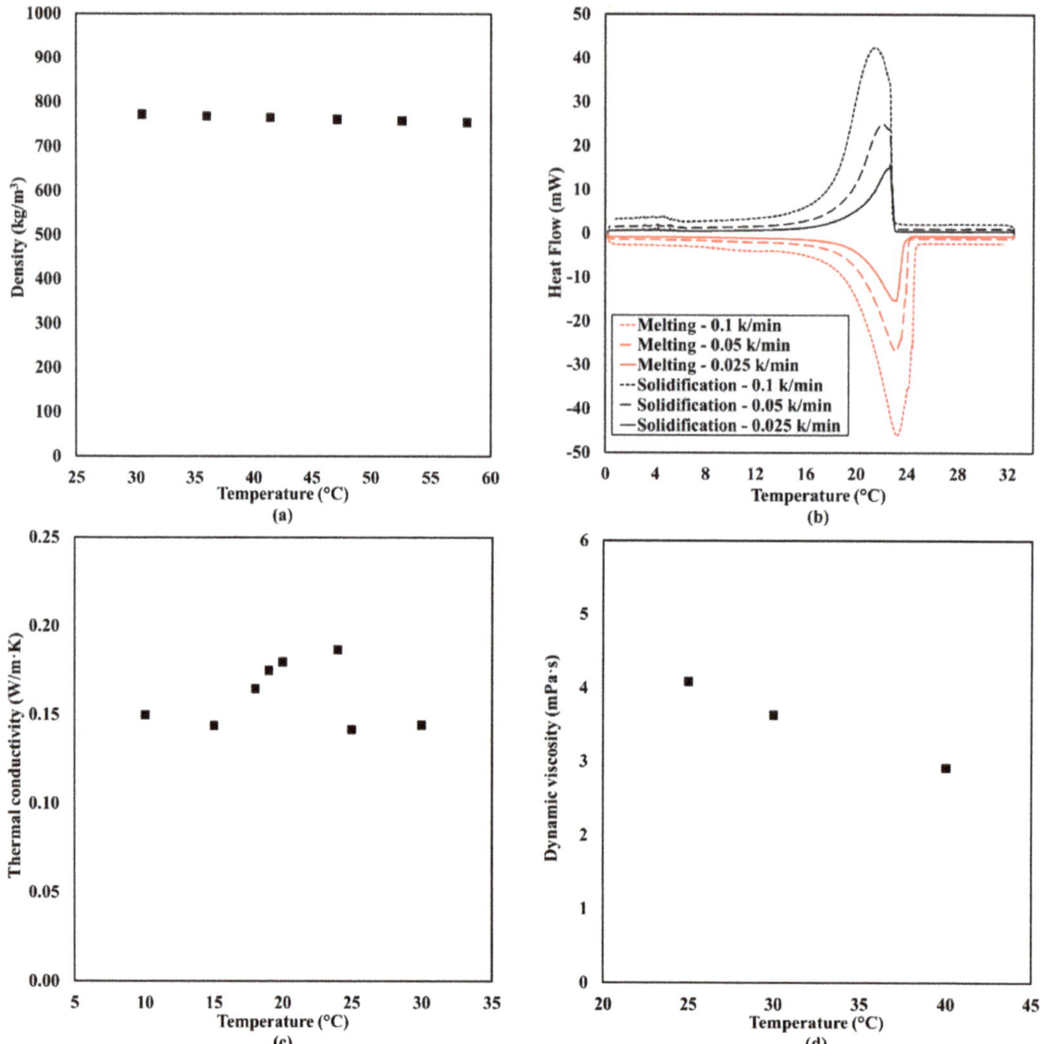

Figure 3. Temperature-dependent thermo-physical properties of RT22: (**a**) liquid density, (**b**) heat flows upon melting/solidification, (**c**) thermal conductivity and (**d**) dynamic viscosity.

Table 3. Measured thermo-physical properties of RT22.

Density ($\frac{kg}{m^3}$)	778 (20 °C)
Specific heat capacity ($\frac{kJ}{kg \cdot K}$)	4.6 (15 °C)–2.4 (30 °C)
Latent heat of phase change ($\frac{kJ}{kg}$)	163.8
ΔT_m (°C)	20.4–23.7
ΔT_f (°C)	20.1–22.8
Thermal conductivity ($\frac{W}{m \cdot K}$)	0.14 (15, 30 °C)
Viscosity (Pa·s)	0.00407 (25 °C)
Thermal expansion coefficient ($\frac{1}{K}$)	0.00084

2.3. Simulation Model

Three-dimensional modeling of the transient phase change is essential to capture the effect of the axial HTF temperature variations since such an effect is missing in two-dimensional modeling. The transient phase change simulations were carried out in three dimensions for the domain specified in Figure 1, using ANSYS Fluent 19.5. In the modeling of dry air flow, the following conditions were considered:

- Incompressible laminar air flow;
- Constant air properties along the channel length;
- Negligible effects of viscous dissipation.

In modeling solid–liquid phase change, the volume change was neglected and natural convection in the constrained melting was modeled via the Boussinesq approximation. The governing mechanism within the phase change is expressed in Equations (1)–(3):

$$\nabla.V = 0 \tag{1}$$

$$\rho\left(\frac{\partial V}{\partial t} + (\nabla.V)V\right) = \rho g - \nabla P + \nabla.(\tau) + S \tag{2}$$

$$\rho\left(\frac{\partial H}{\partial t} + \nabla.(VH)\right) = \nabla.(k\nabla T) \tag{3}$$

An enthalpy–porosity method is used to model the phase change, adding a source term to the momentum equation, defined by the Carmen–Kozeny term shown by Equation (4).

$$S = -\frac{(1-\lambda)^2}{\left(\lambda^3 + \varepsilon\right)}A_{mushy}(V) \tag{4}$$

Here, λ is the porosity factor ranging from zero to one and A_{mushy} is the mushy parameter, which is an arbitrary number ranging from 10^5 to 10^8, representing the convective intensity of the flow within the mushy zone. The lower the mushy numbers, the stronger convective flow is predicted within the mushy zone. In this study, the mushy parameter was set to 10^8 for both the constrained melting and solidification simulations since it was previously shown to give an appropriate fit to experimental data [25–27]. The porosity is incorporated into the sensible and latent energy through Equations (5)–(7).

$$H = h + \lambda L \tag{5}$$

$$h = h_{ref} + \int_{T_{ref}}^{T} C_p dT \tag{6}$$

$$\Delta H = \lambda L \rightarrow \begin{cases} \lambda = 0 & \text{if } T < T_{solidus} \\ \lambda = \frac{T - T_{solidus}}{T_{liquidus} - T_{solidus}} & \text{if } T_{solidus} < T < T_{liquidus} \\ \lambda = 1 & \text{if } T > T_{liquidus} \end{cases} \tag{7}$$

A simple algorithm was used in the coupling of pressure and velocity. The second-order upwind scheme for discretization of momentum and energy equations and the PRESTO! scheme for pressure correction were used. The first-order implicit scheme was employed with fixed time steps for time discretization. The convergence criterion for the continuity and momentum equations was set to 10^{-5} and 10^{-8}, respectively.

A mesh independency study was carried out to select the element size for the final grids of air and PCM. In the beginning, the simulations were carried out for the hydrodynamic laminar air flow in a single channel, and the pressure drops along the channel length in different grids were compared. The element size ranged from 0.4 to 0.1 mm and 4 to 1 mm in the cross-sectional plane and along the channel length, respectively. Figure 4a shows the pressure drop along the length channel-2 for the cases 1–4 with the aforementioned ranging element size. As observed in Figure 4a, the simulated pressure drop values vary insignificantly from case 3 to case 4. The mesh independency analysis proceeded for the PCM domain with the element sizes specified in cases 3 and 4 for the air domain. For the PCM domain, the element sized varied in the cross-sectional plane and in length, and the effect on melt fraction was studied. As shown in Figure 4b, the simulated melt fraction changes slightly with refining the cross-sectional PCM mesh from 1 mm to 0.5 mm. Further reducing the cross-sectional element size in the PCM domain down to 0.3 mm showed an insignificant change in comparison to the element 0.5 mm. Figure 4c shows the influence of decreasing the element size from 3 to 1 mm in the axial direction for the PCM domain, making almost no influence on the melt fraction results. Furthermore, the variation in time-step displays its insignificant effect on the simulations, as shown in Figure 4d. In the end, considering the insignificant changes in the hydrodynamic air flow and melting simulations, the following element sizes were chosen. The final grid elements size, perpendicular to the air flow, were set to 0.4 mm, 0.15 mm and 0.1 mm for the PCM, aluminum and air zones, respectively, as shown in Figure 5, and the grid element size in the axial direction was set to 2 mm. The time step was fixed to 0.1 s in the simulations. In the used meshes, the minimum orthogonal mesh quality in the final grid and the maximum skewness factors were 0.5 and 0.8, respectively.

The simulations with the chosen grid based on the performed mesh independency analysis were performed for the specified configurations in Table 2 with inlet HTF temperature of 30 °C and 15 °C for melting and solidifications, respectively. The initial domain temperature was set to 15 °C and 30°C. The total flow rate of the HTF entering the cuboid was varied within a range of 7–24 L/s. The minimum flow rate, corresponding to a room of 20 m^2, was chosen based on the Swedish regulation of maintaining a minimum flow rate of 0.35 L/m^2·s in the ventilation systems of residential buildings. A parametric study of flow rate with fixed values of 7, 12, 16, 20 and 24 L/s was carried out. As the number of channels was varied, given a constant total air flow rate, the inlet velocity of air flow per channel was calculated and used as the inlet boundary conditions. With the simulated range of flow rate and the varying number of channels from 60 to 100 for channel-1 and 72 to 96 for channel-2, respectively, the Reynolds numbers were within ranges of 250–1440 and 430–1970. The total transient rate of heat transfer is calculated with Equation (8).

$$\dot{Q}(t) = N \rho_{air} V_{air_{in}} A_c \, c_{p_{air}} \, \text{abs}\left[T_{air_{out}}(t) - T_{air_{in}} \right] \tag{8}$$

where N is the number of channels, $\rho_{air} V_{HTF_{in}} A_c$ is the air mass flow rate per channel, $c_{p_{air}}$ is the air specific heat and $T_{air_{out}}(t) - T_{air_{in}}$ is the transient air temperature difference of the outlet and the inlet. The total UA value is calculated as a ratio of the total rate of transferred heat to the logarithmic mean temperature difference (LMTD) between the HTF and the PCM, using Equations (9) and (10).

$$UA = \frac{\dot{Q}(t)}{LMTD(t)} \tag{9}$$

$$LMTD(t) = \frac{\left[T_{air_{in}} - T_{s/m} \right] - \left[T_{air_{out}}(t) - T_{s/m} \right]}{\ln \frac{T_{air_{in}} - T_{s/m}}{T_{air_{out}}(t) - T_{s/m}}} \tag{10}$$

The air heat transfer coefficient is calculated as the ratio of the total heat flux to the logarithmic mean temperature difference between the HTF temperature and the channel wall temperature at the inlet and outlet, as specified in Equations (11) and (12).

$$u_{air}(t) = \frac{\dot{Q}(t)}{A_{air} LMTD_{air}(t)} \tag{11}$$

$$LMTD_{air}(t) = \frac{\left[T_{air_{in}} - T_{w_{in}}(t) \right] - \left[T_{air_{out}}(t) - (T_{w_{out}}(t)) \right]}{\ln \frac{T_{air_{in}} - T_{w_{in}}(t)}{T_{air_{out}}(t) - T_{w_{out}}(t)}} \tag{12}$$

Using the obtained air heat transfer coefficient, the heat transfer coefficient on the PCM side is calculated via Equation (13).

$$u_{PCM}(t) = \frac{1}{A_{PCM}} \left[\frac{1}{UA}(t) - \frac{1}{A_{air}u_{air}}(t) - \frac{W_{thickness}}{k_w A_{air}} \right]^{-1} \tag{13}$$

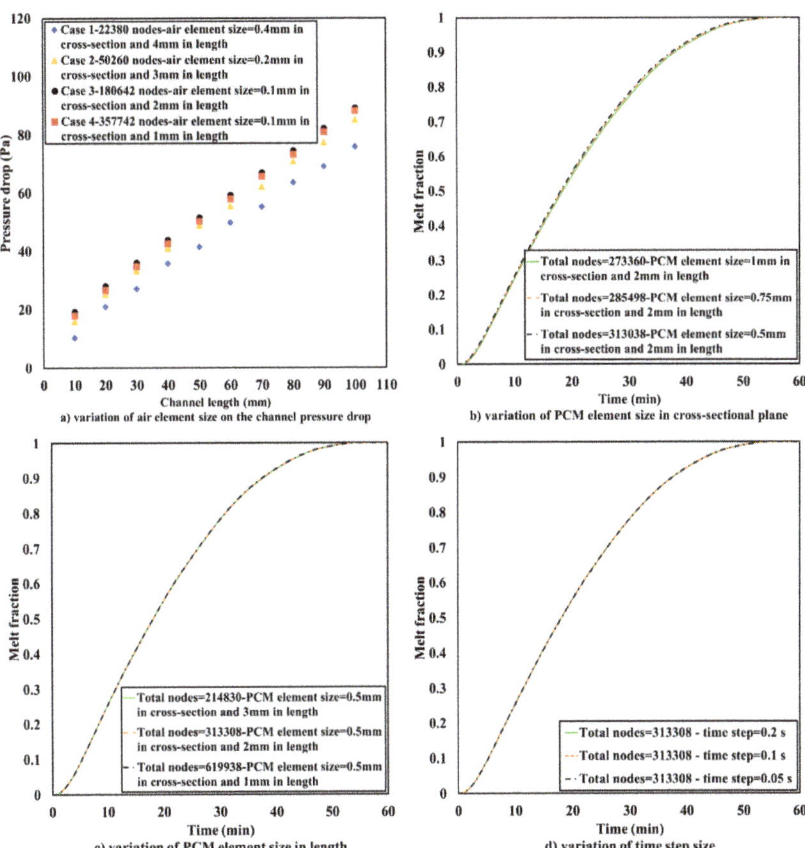

Figure 4. Mesh independency analysis: (**a**) element size variation for pressure drop simulations of air flow in a single channel-2 and V_{air} = 5 m/s, (**b**) element size variation in cross-sectional plane for melt fraction simulations of channel-1 (5 × 16, PF = 82.4%, V_{air} = 10.1 m/s), (**c**) element size variation in length for melt fraction simulations of channel-1 (5 × 16, PF = 82.4%, V_{air} = 10.1 m/s) and (**d**) time-step variation for melt fraction simulations of channel-1 (5 × 16, PF = 82.4%, V_{air} = 10.1 m/s).

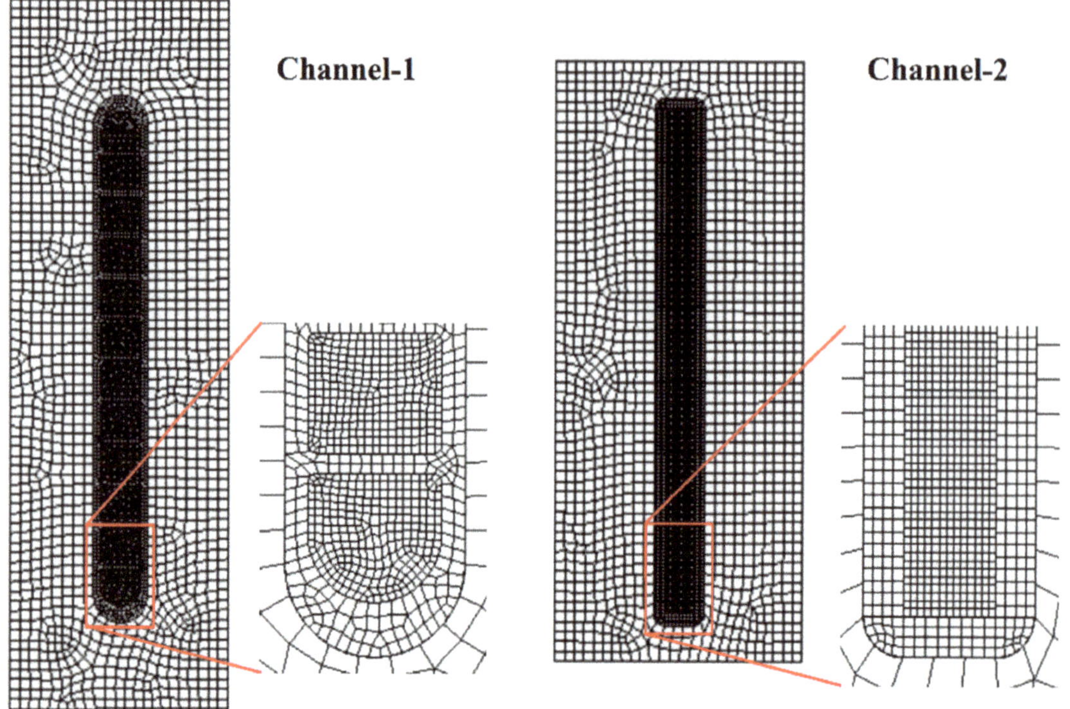

Figure 5. Examples of meshed simulation domains for channel-1 (5 × 16, PF = 82.4%) and channel-2 (6 × 14, PF = 82%).

The total transient power was presented versus relative accumulated extracted/absorbed energy. The relative accumulated energy is defined by Equation (14) as the ratio of accumulated extracted/absorbed energy at each time interval to the total accumulated extracted/absorbed energy.

$$Q_{relative}(t) \; = \; \frac{\sum_{initial}^{t} Q_{accumulated}(t)}{\sum_{initial}^{final} Q_{accumulated}} \times 100 \tag{14}$$

To combine the enhancement in the phase change power and the used fan power and evaluate the system performance, performance evaluation criterion (PEC) is defined via Equation (15) as the ratio of the mean phase change power to the product of total air flow and pressure drop.

$$PEC \; = \; \frac{\dot{Q}_{mean}}{N \, V_{airin} A_c \, \Delta P} \tag{15}$$

3. Results and Discussion

The symmetrical simulated domains of channel-1 and un-finned channel-2, with the ranging number of channels of 60–100 and 72–96, respectively, have different dimensions in width and height. Thus, to have a fair performance assessment of the two channels, the evaluation is made for cases with possibly similar PCM packing factors (PF). The melt fraction contours for the configurations of channel-1 with PF = 82.4% and the bare and finned configurations of channel-2 with PF = 82% and PF = 80.2%, respectively, are shown in Figures 6 and 7. A comparison of melting and solidification processes shows that in both processes, the growing phase change structure is rather symmetric. This indicates the weak presence of the natural convection in melting and the dominancy of conduction in both melting and solidification.

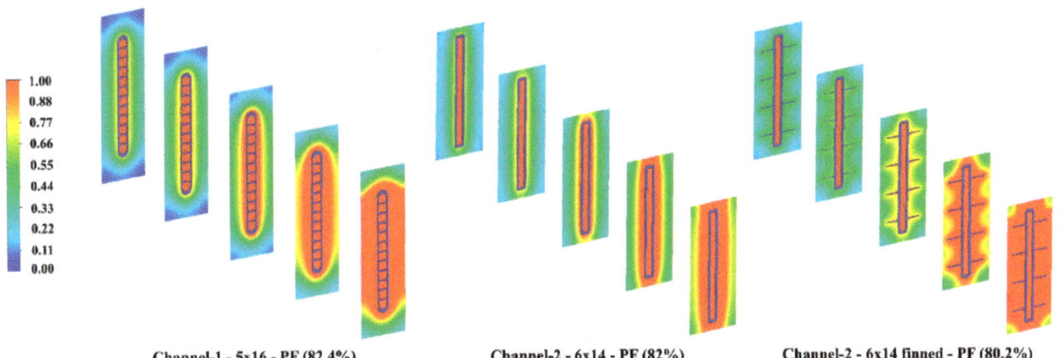

Figure 6. Melt fraction contours within melting for the configurations of channel-1 (PF = 82.4%), channel-2 (PF = 82%) and channel-2-finned (PF = 80.2%).

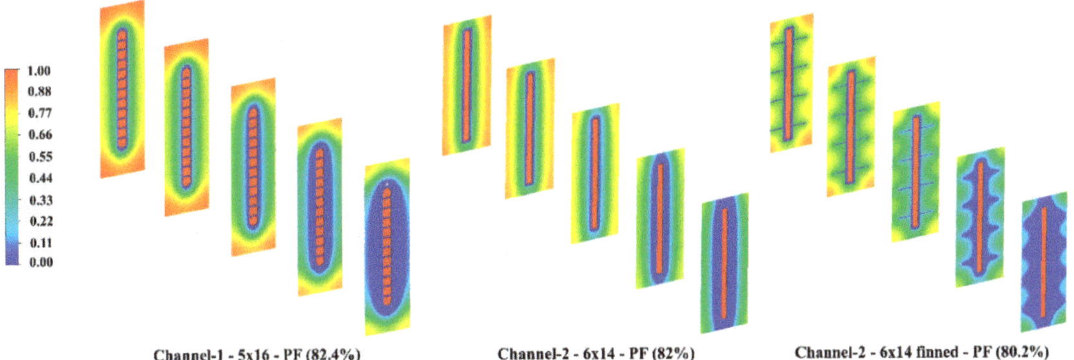

Figure 7. Melt fraction contours within solidification for the configurations of channel-1 (PF = 82.4%), channel-2 (PF = 82%) and channel-2-finned (PF = 80.2%).

The phase change progress for the bare channels is more dominant on the sides of the mini-channels and weaker on the tips. The simulated domain for channel-1 is thinner in width and longer in height than that of channel-2. This results in a rapid phase change progress on the side walls of the former, while the phase change of the PCM around the tips of the channel takes a longer time. In the case of channel-2, a rapid phase change occurs around the tips, given the considerably smaller distance between the tips (3 mm). As the phase change completes around the tips, it progresses afterward on the side walls. In the case of finned channel-2, the fins contribute primarily to the phase change of PCM between the fins as it takes a longer time for the phase change of the PCM around the corners of the domain. The major mechanism at which the fins contribute to the phase change is conduction, given the small 5.3 mm distance between fins.

In all the cases, the phase change process evolves axially. As the phase change gets completed at the inlet of the channels, the phase change front develops in the length of the channels, showing a significant axial variation in the HTF temperature.

Figure 8 shows the transient outlet air temperature of the minimum and maximum simulated flow rates within the phase change time for the configurations of channel-1 (PF = 82.4%) and for the bare and finned configurations of channel-2 (PF = 82% and

PF = 80.2%). As observed, the outlet temperature increases in melting and decreases in solidification steeply at the beginning of the phase change process due to the sensible heat transfer prior to the initiation of phase change along the channel length. As the PCM at the outlet reaches the onset temperature of the phase change and the entire PCM axially adjacent to the channel wall undergoes the phase change process, the outlet temperature varies less steeply, approaching the inlet temperature at the end of the process. The rather large temperature difference between the outlet and inlet for both melting (Tin = 30 °C) and solidification (Tin = 15 °C) indicates the significant axial HTF temperature variations. The axial air temperature change diminishes as the PCM approaches the complete phase change. This could be observed for both the lower and upper bounds of the investigated flow rate range; however, the temperature difference is significantly larger for the former. The outlet air temperature for channel-1 (PF = 82.4%) and the bare channel-2 (PF = 82%) is rather similar throughout the phase change processes, resulting in similar phase change times. The incorporation of fins results in a lower and higher outlet temperature in melting and solidification, respectively, as compared to the other configurations, and a modest reduction in the phase change time. In the finned configuration (PF = 80.2%), with the minimum total air flow rate of 7 L/s, the phase change time in melting and solidification, respectively, has decreased by 16% and 10%, as compared to the bare channel-2 configuration. On the contrary, the influence of increasing the air flow rate is rather considerable for all the shown cases. For a packing factor range of 86.8% to 77.4%, varying the air flow rate from 7 to 24 L/s reduces the melting and solidification time by 41% to 53% and 39% to 53%, respectively.

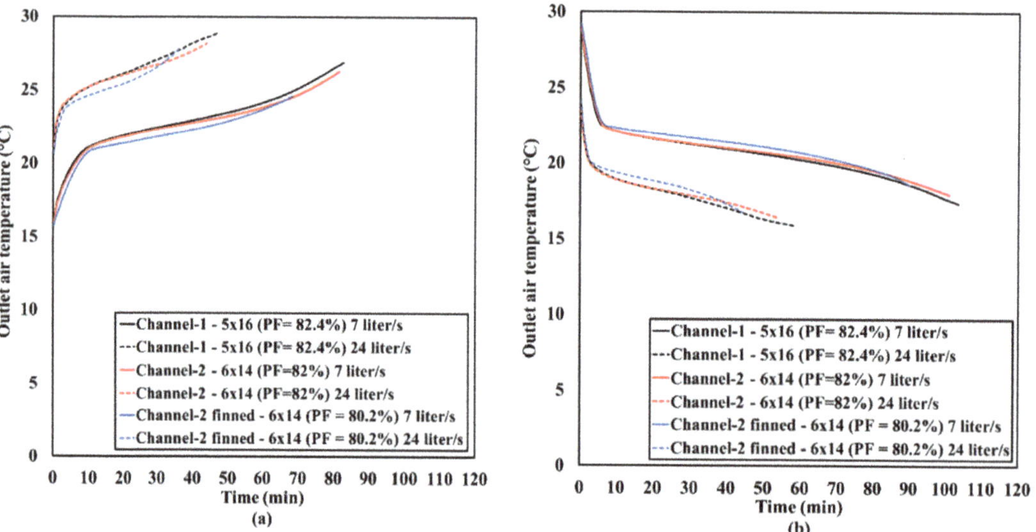

Figure 8. Transient outlet air temperature of flow rates of 7 and 24 L/s: (**a**) melting-Tin = 30 °C, (**b**) solidification-Tin = 15 °C.

The effect of air flow rate increase on the transient power during phase change as a function of relative accumulated energy is shown in Figure 9. At the minimum air flow rate, the transient power is rather similar for the shown cases of channel-1 and channel-2, and the gain in the finned channel-2 is insignificant. Increasing the flow rate to the upper bound of 24 L/s increases the transient power considerably, indicating the resistance on the air side is not negligible. However, the enhancement effect of the increased air flow rate is accompanied by a significant increase in the pressure drop. A comparison of the channels at the maximum flow rate shows that the channel-1 case gives a slightly higher transient power up to about 70% of the relative stored/extracted energy, as compared to the channel-2 case. For the rest of the process, the channel-2 case has a higher power. The rapid reduction in the power of channel-1 at the end of the processes is attributed to the completion of the phase change adjacent to the side walls and the relatively weaker heat transfer around the tips of channel-1. In the case of channel-2, the phase change is ongoing adjacent to the side walls throughout the entire process. The finned channel-2 case shows a higher rate of heat transfer in both melting and solidification processes than the bare channel-2, despite having a lower mass of PCM with about 1.8% lower packing factor.

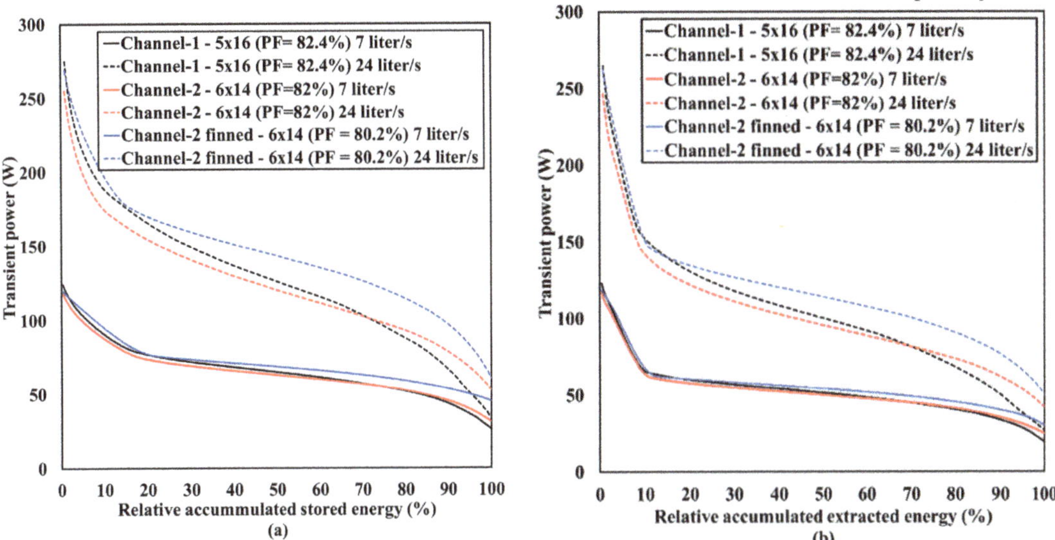

Figure 9. Transient power as a function of relative accumulated energy flow rates of 7 and 24 L/s: (**a**) melting-Tin = 30 °C, (**b**) solidification-Tin = 15 °C.

To be able to evaluate and compare the designs with a varying number of channels and packing factors, normalized mean power (represented by the mean UA-value) as an indicator of the thermal performance was used in the following. The mean UA-values are calculated based on 98% completion of the phase change. Figure 10 shows the dependency of UA values as a function of the PCM packing factor for the total range of the simulated air flow rate. The reduction in the packing factor indicates the increase in the number of channels as the vertical data points for a given packing factor designates the variation in the air flow rate. Figure 10 shows that as the PCM packing factor reduces, the mean UA value increases for all the simulated configurations in both melting and solidification processes.

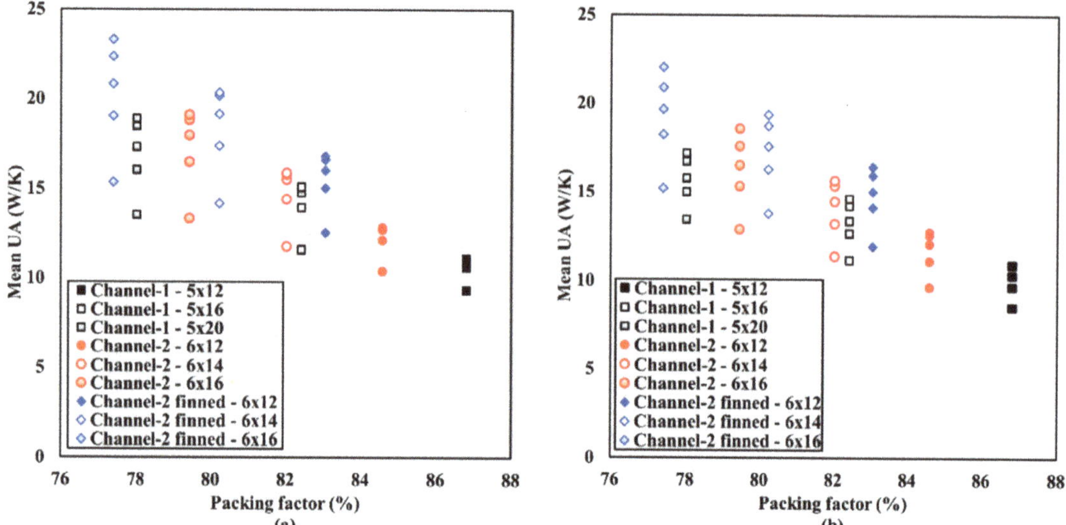

Figure 10. Mean UA values for the range of the simulated flow rate as functions of the PCM packing factor: (**a**) melting-Tin = 30 °C, (**b**) solidification-Tin = 15 °C.

For a constant air flow rate, the increase in the number of channels increases the mean power of phase change. As the number of channels is increased or fins are used (in the case of channel-2), the PCM packing factor reduces, resulting in a reduction in the latent storage capacity. Given a constant total flow rate of the HTF entering the cuboid, the increase in the number of channels reduces the flow rate per channel, resulting in a reduction in the heat transfer coefficient on the air side. However, the effect of the extended heat transfer surface as the number of channels increases is dominant over the reduction in the internal heat transfer coefficient, eventually increasing the total UA value and improving the thermal performance.

Increasing the flow rate improves the normalized mean power for all the cases, particularly for the enhanced case of finned channel-2. For instance, for the configurations of channel-1 with PF = 82.4%, channel-2 with PF = 82% and finned channel-2 with PF = 80.2%, increasing the total flow rate of the HTF from 7 L/s to 24 L/s increases the mean UA values of melting by 30%, 35% and 44%, respectively.

The interesting point is that adding fins to the cases of channel-2 instead of increasing the number of channels can result in higher mean UA values, attributed to the higher air flow per channel in the former case. As an example, given the flow rate range of 7–24 L/s, adding fins to the case with 72 channels (PF = 83%) increases the UA mean value of melting by 21–31%, as compared to the un-finned case with 72 channels (PF = 84.6%). Whereas increasing the number of un-finned channels from 72 (PF = 84.6%) to 84 (PF = 80.2%) enhances the mean UA value of melting by 13–24%. In total, the relative enhancements in the mean UA values of the finned channel-2 cases, as compared to the un-finned cases, lie within 15–31% for the entire range of simulated flow rates and packing factors.

The difference in the air heat transfer coefficient of the two channels has an insignificant effect on their thermal performance within the phase change. Figure 11a shows the air heat transfer coefficients of channel-1 and bare channel-2 averaged over melting as functions of the total air flow rate. Channel-2 has a modestly higher heat transfer coefficient on the air side within the simulated air flow rate, despite its higher hydraulic diameter as compared to channel-1. This is due to the higher aspect ratio of the flow passage. Even though there is a considerable distinction in the air heat transfer coefficient between channel-1 and channel-2, as shown in Figure 11a, the heat transfer coefficient on the PCM side remains unaffected. Figure 11b shows the heat transfer coefficient on the PCM side for channel-1 and channel-2 cases with PF = 82.4% and PF = 82%, respectively, for flow rates of 7 and 24 L/s as functions of melt fraction in the melting process. The PCM heat transfer values drop quickly from the high values at the beginning of the process to levels below the heat transfer coefficients of air, shifting the thermal resistance from the air side to the PCM side.

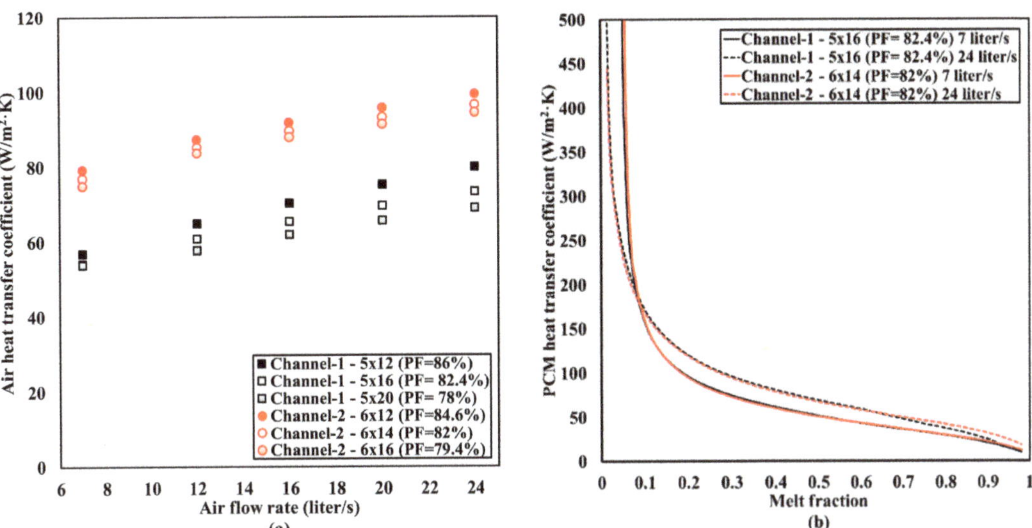

Figure 11. (**a**) mean air heat transfer coefficient during melting, (**b**) transient heat transfer coefficient on the PCM side during melting.

As observed, the PCM heat transfer coefficients of channel-1 and channel-2 relatively overlap throughout the process. A clear distinction is seen from the lower bound of the air flow rate to the upper bound, implying the PCM heat transfer coefficient could be a function of the internal air flow rate.

The major disadvantage of extending the surface area and UA value through fin incorporation rather than increasing the number of channels is the higher flow rate per channel in the former, given a fixed total flow rate, unfavorably resulting in a higher pressure drop. Figure 12 shows the pressure drop for channel-1 and channel-2 configurations. As seen, with high packing factors and a lower number of channels, the air flow entering each channel increases. In the cases with high packing factors, as the total flow rate increases, the pressure drop increases significantly while the internal air heat transfer coefficient increases modestly. Increasing the number of channels reduces the pressure drop considerably, particularly at high total air flow rates. As an example, at the flow rate of 24 L/s, the pressure drop for channel-1 reduces by 31% and 47% as the number of channels increases from 60 to 80 and 100, and for channel-2 it reduces by 17% and 30% as the number of channels increases from 72 to 84 and 96, respectively. Interesting to point out is the relatively lower pressure drop of channel-2 than channel-1. This is credited mainly to the higher hydraulic diameter of channel-2.

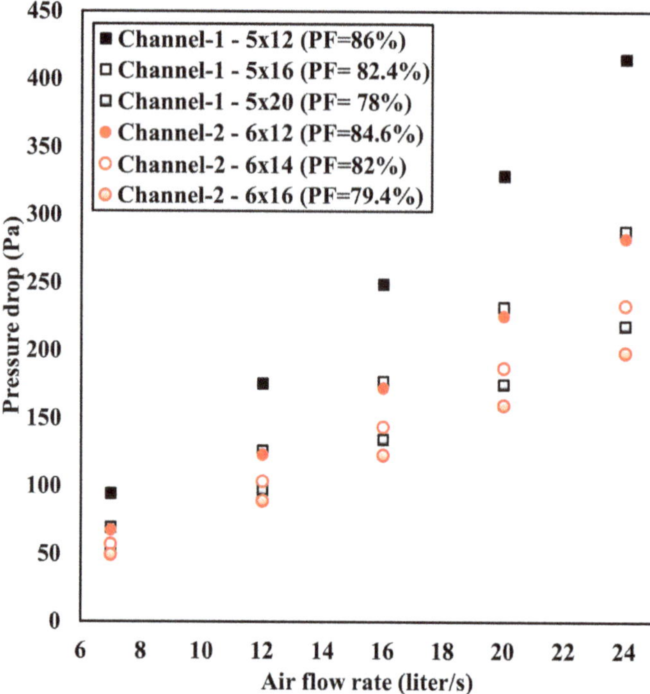

Figure 12. Pressure drop as a function of total air flow rate.

Having a low pressure drop and, consequently, a low fan power as possible is a crucial matter in employing mini-channels in LTES components. This contributes to achieving a higher gain in terms of charging/discharging thermal energy in exchange for fan power. The performance evaluation criterion (PEC) of the compact component defined as the ratio of the mean phase change power to the fan work (Equation (15)) is shown in Figure 13 for the simulated configurations as a function of the total air flow rate. Considering the increase in the total air flow rate, the PEC for both melting and solidification reduces dramatically, indicating a more significant increase in the pressure drop compared to the modest gain in the mean phase change power. The same trend is observed as the number of channels reduces and the packing factor increases. Given a constant total flow rate, channel-2 is superior to channel-1 due to both the relatively higher mean phase change power and the considerably lower pressure drop. In addition, in the case of finned channel-2, fins were able to increase the PEC modestly. Since the pressure drops in the finned channel-2 cases are the same as the un-finned channel-2 cases, the gain in the PEC comes from the modest increase in the mean phase change power.

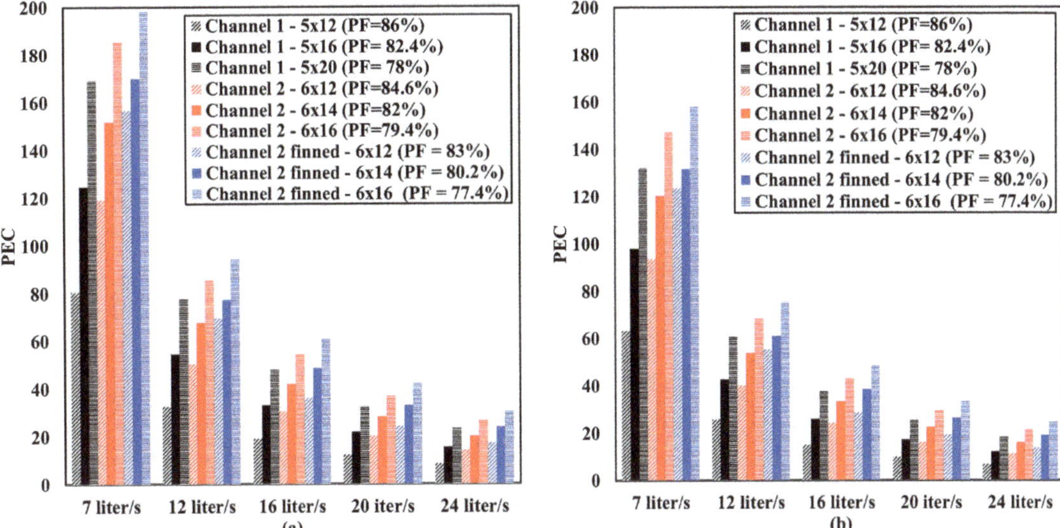

Figure 13. PEC of the compact component as a function of total air flow rate: (**a**) melting—Tin = 30 °C, (**b**) solidification—Tin = 15 °C.

4. Conclusions

This incorporation of rectangular mini-channels with internal air passages into an LTES component was studied, considering two types of configurations. Three-dimensional phase change simulations of RT22 in a cuboid of $0.15 \times 0.15 \times 0.1$ m^3, with vertically oriented channels, were carried out. The channels have different hydraulic diameters and aspect ratios of the internal flow passages while having almost similar volume and external heat transfer surfaces. In channel-2, fins were also investigated to enhance thermal performance.

The simulation results show that the thermal performance of the system is enhanced as the number of channels increases in both the channel-1 and channel-2 cases. The mean UA value increases as the number of channels increases and the PCM packing factor decreases. The enhancement is majorly attributed to the increase in the external surface area with a higher number of channels, while the air flow rate per channel decreases given a constant total air flow rate entering the cuboid.

Increasing the flow rate improves the mean UA value in both melting and solidification. For instance, for channel-1 with PF = 82.4%, channel-2 with PF = 82% and finned channel-2 with PF = 80.2%, the mean UA value of melting increases by 30%, 35% and 44% as the total flow rate of the HTF increases from 7 L/s to 24 L/s, respectively.

The comparison of channel-1 and channel-2 with similar packing factors shows that the difference in the air heat transfer coefficient has an insignificant effect on the mean phase change powers and the mean UA values. For instance, the air heat transfer coefficients of channel-2 with PF = 82.0% are about 30–40% higher than those of channel-1 with PF = 82.4%, for the simulated total flow rate of 7–24 L/s. However, the enhancement in the mean UA values of the former lies within 1–5%, as compared to the latter, credited to its minor higher external surface area. Adding fins to the channel-2 cases, doubling the heat transfer area on the PCM side, can increase the mean UA values by 15–31% for the entire ranges of simulated flow rates and packing factors.

Another benefit with channel-2 is the considerably lower pressure drop as compared to channel-1. This is due to its higher hydraulic diameter, particularly with high total flow rates and high packing factors. Given a constant flow rate, the channel-2 configurations have higher PEC than the channel-1 configurations, achieving a higher gain in stored/extracted thermal energy in exchange for the used fan power.

The results show that the mini-channels could be optimized to have a higher performance in heat transfer with a lower pressure drop. The metal 3D printing method, as an alternative method in manufacturing conceptual designs, could be used in implementing innovative features. This includes optimization of the hydraulic diameter and the aspect ratio of the HTF passages, maximizing the area-to-volume ratio of the channels and manufacturing innovative fin configurations to enhance the heat transfer on the PCM side. Finally, optimization, manufacturing and experimental testing of the mini-channels under laboratory conditions could be the next steps paving the commercialization of such products and their usage on a large scale.

Author Contributions: Conceptualization, A.A.; methodology, A.A.; software, A.A.; investigation, A.A.; data curation, A.A.; writing—original draft preparation, A.A.; writing—review and editing, J.N.C. and V.M.; visualization, A.A.; supervision, J.N.C. and V.M.; project administration, V.M.; funding acquisition, V.M. All authors have read and agreed to the published version of the manuscript.

Funding: This work was supported and funded by the Swedish Energy Agency (P48283-1).

Institutional Review Board Statement: Not applicable.

Informed Consent Statement: Informed consent was obtained from all subjects involved in the study.

Data Availability Statement: The data presented in this study are available on request from the corresponding author.

Acknowledgments: Authors would like to thank Hydro aluminum for providing the aluminum profiles. The computations were enabled by resources provided by the Swedish National Infrastructure for Computing (SNIC) at the PDC Center for High Performance Computing, KTH Royal Institute of Technology, partially funded by the Swedish Research Council through grant agreement no. 2016-07213.

Conflicts of Interest: The authors declare no conflict of interest.

Nomenclature

A_c	cross sectional area (mm^2)
As	heat transfer surface area (mm^2)
A_{mushy}	mushy parameter
AR	aspect ratio
c_p	specific heat ($\frac{kJ}{kg \cdot K}$)
d_h	hydraulic diameter (mm)
g	gravity ($\frac{m}{s^2}$)
k	thermal conductivity ($\frac{W}{m \cdot K}$)
h	sensible enthalpy ($\frac{kJ}{kg}$)
H	total enthalpy ($\frac{kJ}{kg}$)
L	latent heat of phase change ($\frac{kJ}{kg}$)
N	number of channels
$\dot{Q}$	power (W)
Q	energy (kJ)
P	pressure (Pa)
Re	Reynolds number [$\frac{\rho V d_h}{\mu}$]
S	source term
t	time (s)
T	temperature (°C)
UA	total heat transfer unit ($\frac{W}{K}$)
U	heat transfer coefficient ($\frac{W}{m^2 \cdot K}$)
V	velocity ($\frac{m}{s}$)
Vol	volume (m^3)

Abbreviations

HTF	heat transfer fluid
LMTD	logarithmic mean temperature difference
LTES	latent thermal energy storage
PCM	phase change material
PEC	performance evaluation criterion

Subscripts

avg	average
in	inlet
m	melting
out	outlet
s	solid
w	wall

Greek Letters

α	Thermal diffusivity $(\frac{K}{\rho c_p})$
β	expansion coefficient $(\frac{1}{K})$
Δ	Difference
ε	small number
λ	porosity factor
μ	viscosity (Pa·s)
ρ	density $(\frac{kg}{m^3})$

References

1. Iten, M.; Liu, S.; Shukla, A. A review on the air-PCM-TES application for free cooling and heating in the buildings. *Renew. Sustain. Energy Rev.* **2016**, *61*, 175–186. [CrossRef]
2. Khan, Z.; Khan, Z.; Ghafoor, A. A review of performance enhancement of PCM based latent heat storage system within the context of materials, thermal stability and compatibility. *Energy Convers. Manag.* **2016**, *115*, 132–158. [CrossRef]
3. Zalba, B.; Marín, J.M.; Cabeza, L.F.; Mehling, H. Review on thermal energy storage with phase change: Materials, heat transfer analysis and applications. *Appl. Therm Eng.* **2003**, *23*, 251–283. [CrossRef]
4. Xu, T.; Chiu, J.N.; Palm, B.; Sawalha, S. Experimental investigation on cylindrically macro-encapsulated latent heat storage for space heating applications. *Energy Convers. Manag.* **2019**, *182*, 166–177. [CrossRef]
5. Wang, X.; Liu, J.; Zhang, Y.; Di, H.; Jiang, Y. Experimental research on a kind of novel high temperature phase change storage heater. *Energy Convers. Manag.* **2006**, *47*, 2211–2222. [CrossRef]
6. Dolado, P.; Lazaro, A.; Marin, J.M.; Zalba, B. Characterization of melting and solidification in a real scale PCM-air heat exchanger: Numerical model and experimental validation. *Energy Convers. Manag.* **2011**, *52*, 1890–1907. [CrossRef]
7. Vakilaltojjar, S.M.; Saman, W. Analysis and modelling of a phase change storage system for air conditioning applications. *Appl. Therm. Eng.* **2001**, *21*, 249–263. [CrossRef]
8. Marin, J.M.; Zalba, B.; Cabeza, L.F.; Mehling, H. Improvement of a thermal energy storage using plates with paraffin–graphite composite. *Int. J. Heat Mass Transf.* **2005**, *48*, 2561–2570. [CrossRef]
9. Zalba, B.; Marín, J.M.; Cabeza, L.F.; Mehling, H. Free-cooling of buildings with phase change materials. *Int. J. Refrig.* **2004**, *27*, 839–849. [CrossRef]
10. Herbinger, F.; Bhouri, M.; Groulx, D. Investigation of heat transfer inside a PCM-air heat exchanger: A numerical parametric study. *Heat Mass Transf.* **2018**, *54*, 2433–2442. [CrossRef]
11. Kandlikar, S.; Garimella, S.; Li, D.; Colin, S.; King, M.R. *Heat Transfer and Fluid Flow in Minichannels and Microchannels*; Elsevier: Amsterdam, The Netherlands, 2005.
12. Lee, P.-S.; Garimella, S.V. Saturated flow boiling heat transfer and pressure drop in silicon microchannel arrays. *Int. J. Heat Mass Transf.* **2008**, *51*, 789–806. [CrossRef]
13. Jagirdar, M.; Lee, P.S. Quasi-steady and transient study of heat transfer during sub-cooled flow boiling in a small aspect ratio microchannel. *Int. J. Multiph. Flow* **2020**, *133*, 103446. [CrossRef]
14. Zeng, S.; Kanargi, B.; Lee, P.S. Experimental and numerical investigation of a mini channel forced air heat sink designed by topology optimization. *Int. J. Heat Mass Transf.* **2018**, *121*, 663–679. [CrossRef]
15. Wibulswas, P. Laminar-Flow Heat-Transfer in non-Circular Ducts. Doctoral Dissertation, University College London, London, UK, 1966.

16. Kaew-On, J.; Sakamatapan, K.; Wongwises, S. Flow boiling heat transfer of R134a in the multiport minichannel heat exchangers. *Exp. Therm. Fluid Sci.* **2011**, *35*, 364–374. [CrossRef]
17. Robles, A.; Duong, V.; Martin, A.J.; Guadarrama, J.L.; Diaz, G. Aluminum minichannel solar water heater performance under year-round weather conditions. *Sol. Energy* **2014**, *110*, 356–364. [CrossRef]
18. Tang, D.; Fang, W.; Fan, X.; Li, D.; Peng, Y. Effect of die design in microchannel tube extrusion. *Procedia Eng.* **2014**, *81*, 628–633. [CrossRef]
19. Duong, V.T. Minichannel-Tube Solar Thermal Collectors for Low to Medium Temperature Applications. Ph. D. Thesis, University of California Merced, Merced, CA, USA, 2015.
20. Rastan, H. Investigation of the Heat Transfer of Enhanced Additively Manufactured Minichannel Heat Exchangers. M.Sc. Thesis, KTH Royal Institute of Technology, Stockholm, Sweden, 2019.
21. Rastan, H.; Abdi, A.; Hamawandi, B.; Ignatowicz, M.; Meyer, J.P.; Palm, B. Heat transfer study of enhanced additively manufactured minichannel heat exchangers. *Int. J. Heat Mass Transf.* **2020**, *161*, 120271. [CrossRef]
22. Voller, V.R.; Prakash, C. A fixed grid numerical modelling methodology for convection-diffusion mushy region phase-change problems. *Int. J. Heat Mass Transf.* **1987**, *30*, 1709–1719. [CrossRef]
23. Voller, V.R.; Brent, A.D.; Prakash, C. The modelling of heat, mass and solute transport in solidification systems. *Int. J. Heat Mass Transf.* **1989**, *32*, 1719–1731. [CrossRef]
24. Abdi, A.; Ignatowicz, M.; Gunasekara, S.N.; Chiu, J.N.W.; Martin, V. Experimental investigation of thermo-physical properties of n-octadecane and n-eicosane. *Int. J. Heat Mass Transf.* **2020**, *161*, 120285. [CrossRef]
25. Shmueli, H.; Ziskind, G.; Letan, R. Melting in a vertical cylindrical tube: Numerical investigation and comparison with experiments. *Int. J. Heat Mass Transf.* **2010**, *53*, 4082–4091. [CrossRef]
26. Assis, E.; Ziskind, G.; Letan, R. Numerical and Experimental Study of Solidification in a Spherical Shell. *J. Heat Transf.* **2009**, *131*, 024502. [CrossRef]
27. Assis, E.; Katsman, L.; Ziskind, G.; Letan, R. Numerical and experimental study of melting in a spherical shell. *Int. J. Heat Mass Transf.* **2007**, *50*, 1790–1804. [CrossRef]

applied
sciences

Article

Stability Study of Erythritol as Phase Change Material for Medium Temperature Thermal Applications

Mayra Paulina Alferez Luna *, Hannah Neumann and Stefan Gschwander

Fraunhofer Institute for Solar Energy Systems, Heidenhofstr. 2, 79110 Freiburg, Germany;
hannah.neumann@ise.fraunhofer.de (H.N.); Stefan.gschwander@ise.fraunhofer.de (S.G.)
* Correspondence: mayra.alferez.luna@ise.fraunhofer.de; Tel.: +49-(761)-4588-2016; Fax: +49-761-4588-9000

Abstract: Sugar alcohols belong to a promising category of organic phase change materials (PCM) because of their high latent heat and density compared to other PCM. However, some sugar alcohols have shown latent heat degradation when heated above their melting temperature. Most of the available studies report the structural changes of erythritol during cycling rather than its thermal stability at constant temperature. This study aimed to assess the effect of thermal treatment on erythritol thermal, chemical and physical properties, as well as to find means to enhance its thermal stability. Erythritol and its mixtures with antioxidant were heated and maintained at different temperatures above its melting point. Erythritol was analyzed before and after thermal treatment via Fourier-transform infrared spectroscopy and differential scanning calorimetry. It was suggested that the degradation of latent heat follows a first order reaction. Mixtures of erythritol with antioxidant had a lower degradation rate compared to pure erythritol under air. Sample browning was observed along the heating treatment of mainly pure erythritol. Antioxidant was found to help to reduce erythritol degradation. No chemical composition changes were detected in samples under argon atmosphere and overall good thermal stability was found throughout the testing period.

Keywords: phase change material; sugar alcohol; erythritol; latent heat storage; thermal stability; degradation kinetics

Citation: Alferez Luna, M.P.; Neumann, H.; Gschwander, S. Stability Study of Erythritol as Phase Change Material for Medium Temperature Thermal Applications. *Appl. Sci.* **2021**, *11*, 5448. https://doi.org/10.3390/app11125448

Academic Editor: Ioannis Kartsonakis

Received: 10 May 2021
Accepted: 9 June 2021
Published: 11 June 2021

Publisher's Note: MDPI stays neutral with regard to jurisdictional claims in published maps and institutional affiliations.

1. Introduction

An efficient use of renewable energy resources, such as solar energy, is increasingly being considered as a promising solution to global warming. However, solar energy is intermittent in nature and its intensity depends on the hour of the day and local weather conditions [1,2].

It is necessary to establish thermal energy storage (TES) technologies in order to steadily utilize the unused thermal energy. The main role of energy storage systems is to reduce the time or rate mismatch between energy supply and energy demand [3]. In Europe, it has been estimated that around 1.4 million GWh per year can be saved and 400 million tons of CO_2 emissions avoided in buildings and in industrial sectors by more extensive use of heat and cold storage [4].

Sensible and latent heat storage are considered as the basic types of TES techniques. Sensible heat storage (SHS) is based on the temperature change in the material. SHS systems use the heat capacity and the change in temperature of the storage medium during the charging/discharging process. Therefore, the temperature of the storage material increases when energy is absorbed and decreases when energy is released [5]. Latent heat storage (LHS) is based on the heat absorption and release when the storage medium undergoes a phase change. A solid-liquid phase change via melting and solidification can store large amounts of thermal energy. Usually, a small volume change ($\leq 10\%$) occurs upon melting. Therefore, the pressure is not changed significantly and, hence, melting and solidification of the storage material proceed at a constant temperature. Upon melting, the storage material keeps its temperature constant at the melting temperature, which is also called temperature of transition [6]. Stored latent heat during the phase change process is calculated from the enthalpy difference (ΔH) between the solid and liquid phase, which is called melting enthalpy or heat of fusion.

Phase change material (PCM) can be used to store thermal energy or to control the temperature swings within a specific temperature range. The operating principle of PCM is the following: as the temperature rises and reaches its melting temperature, PCM absorbs heat in an endothermic process and changes phase from solid to liquid. Similarly, as the temperature drops under its melting temperature, PCM releases heat in an exothermic process and returns to its solid phase [7]. According to its properties, PCMs have several application possibilities. A PCM filled window system was proposed by Ismail and Henriquez [8] taking advantage of the translucent and thermal properties of the selected PCM. Some PCMs are highly transparent for the visible part of solar radiation whereas the infrared part is absorbed within the PCM, which allows both the advantages of a daylighting element and of energy storage to be achieved [9]. In addition, coconut fat recovered from underused feedstocks have been used as bio-based PCM for its application in building envelopes due to its good physical, chemical and thermal properties [10].

Sugar alcohols (SA), also called polyols [11], are attractive bio-based materials to be used as PCM. Additionally, as by-products from the food industry, SA are non-corrosive, non-toxic and environmentally friendly [12]. They have great potential in many TES applications due to their high mass and volume-specific melting enthalpies compared to other organic PCM [13]. SA have melting temperatures ranging between 90 and 200 °C. However, some SA are not stable at higher temperatures [6].

Erythritol is a promising organic PCM for recovering solar energy and industrial waste heat at temperatures up to 160 °C [14]. Erythritol is a material emerging as PCM due to its attractive TES properties. Erythritol's melting point varies in literature [11,14–16] between 117 °C and 119 °C according to its purity percent and manufacturer. In addition, it has a high latent heat of 340 J/g, almost equivalent to that of the ice to liquid water phase change.

Kaizawa et al. [14] studied the thermophysical properties of erythritol using differential scanning calorimetry (DSC), thermogravimetry-differential thermal analysis (TG-DTA) and a heat storage tube. Palomo del Barrio et al. [11] carried out an experimental characterization of erythritol by measuring their physical properties such as specific heat and thermal conductivity, among others. The suitability of erythritol as PCM was examined by Kakiuchi et al. [15] according to its thermodynamic, kinetic and chemical properties. Zhang et al. [17] found that the crystal growth speed of some SA is highly dependent on temperature, whereas Lopes Jesus et al. [18] claimed that the solid erythritol obtained from the melt has different degrees of crystallization depending on the used cooling rate. Some eutectic mixtures have also been studied, such as erythritol-xylitol or erythritol-sorbitol [19], and erythritol-based composites with the addition of expanded graphite and carbon nanotubes were also evaluated by Guo et al. [20]. However, all these studies addressed thermal cycling tests rather than erythritol thermal stability at constant temperature, which would allow a degradation rate to be determined under conditions closer to real operation. On the other hand, Nomura et al. [16] investigated the heat release performance of a direct-contact heat exchanger using erythritol as PCM along with a heat-transfer oil. The suitability and appropriate parameters for charging and discharging erythritol to power a LiBr/H_2O absorption cooling system were assessed by Agyenim et al. [21]. Kaizawa et al. [22] studied the thermal and flow behaviors in a trans-heat container filled with erythritol as part of a waste heat transportation system. Nevertheless, the conclusions made by these authors are ambiguous about the performance of erythritol as PCM.

Therefore, this study aims to investigate the effect of thermal treatment on the erythritol thermal properties at constant temperatures above its melting point. The experiments are also conducted under an inert atmosphere and an antioxidant is added to the samples. Degradation rates at different temperatures and conditions (under air/inert atmosphere and with/without antioxidant) are calculated. In addition, degradation kinetics with respect to its latent heat are evaluated. A mechanism of oxidation of erythritol under air is proposed, as well as the possible underlying mechanism on the stability enhancement of erythritol by antioxidant. The reaction pathway of thermal dehydration of erythritol and the likely relationship between dehydration and oxidation products with the sample browning are suggested

2. Materials and Methods

2.1. Materials

ZeroseTM Erythritol with a purity of 99.5% was purchased from Cargill and used without further treatment. Irganox 1010$^{®}$ from BASF [23] was used as antioxidant. Irganox 1010$^{®}$ is a phenolic antioxidant which has been used to stabilize organic polymers against thermal oxidative degradation [24,25].

2.2. Sample Preparation

Samples were prepared in a mortar by manually grinding and mixing erythritol with 5 wt.% of antioxidant. An amount of 220 ± 0.4 mg of each sample was filled into the glass vials (1.5 mL, ND11) and then sealed using a PTFE/silicon septum and an aluminum crimp cap. In order to create an anoxic environment inside the vials, some samples were purged with argon for 10 min by using two cannulas as inlet and outlet. Samples that did not get further treatment after the sealing process were used to assess the degradation under air atmosphere. Pure erythritol samples were also used as blanks for each experiment. All samples were weighed after sealing them using a Secura$^{®}$ Analytical Balance from Sartorius (224-1S) with an uncertainty of ±0.1 mg.

The glass vials filled with the samples were placed in aluminum blocks according to their atmosphere (air and argon). Each of these blocks had an aluminum lid, which was attached with four screws (Figure 1). For the experiments under argon atmosphere, a gas inlet and outlet of the aluminum blocks was connected with a plastic hose to nitrogen stream, which flowed through the block continuously (Figure 1). A continuous flow rate of 30 cm^3/min of nitrogen was used then to reduce the traces of oxygen that may have remained between the block cavities and the glass vials, as well as to prevent it from entering the vials.

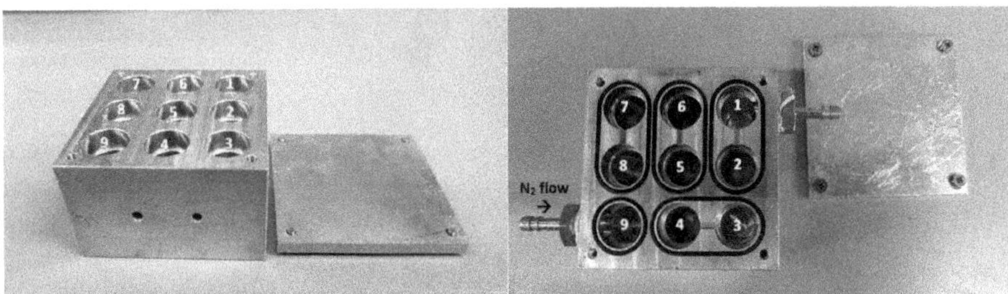

Figure 1. Left: aluminum block used for tests under air. Right: aluminum block with gas inlet and outlet for tests under argon atmosphere.

2.3. Thermal Treatment

Once the aluminum blocks were loaded with the sample vials, they were heated and maintained at a constant temperature of 121 °C, 131 °C and 141 °C ± 1 °C, respectively. After 10 ± 5 h, one vial after the other was taken out of the aluminum block. Samples of pure erythritol under air and under argon with antioxidant at 141 °C had a shorter heating time (8 ± 4 h) due to the technical capacity in the lab. This method allowed all batch samples (9 samples per block) to be exposed to the heating temperature for different periods of time, and hence, changes in the melting enthalpy could be assessed while increasing the thermal treatment time. Thermal degradation tests were performed for about a maximum of 100 h. In addition, samples of pure erythritol under argon atmosphere and with 5 wt.% antioxidant under air, both at 141 °C, were assessed over a period of 935 h. Table 1 summarizes the performed experiments.

Table 1. Summary of the performed experiments.

Atmosphere	Sample	Temperature [°C]	Thermal Treatment Time	Test Duration
Air	Erythritol	121	100 h	Short-term-test
	Erythritol/Antiox		100 h	Short-term-test
	Erythritol	131	100 h	Short-term-test
	Erythritol/Antiox		100 h	Short-term-test
	Erythritol	141	78 h	Short-term-test
	Erythritol/Antiox		100 h	Short-term-test
Argon	Erythritol	121	100 h	Short-term-test
	Erythritol/Antiox		100 h	Short-term-test
	Erythritol	131	100 h	Short-term-test
	Erythritol/Antiox		100 h	Short-term-test
	Erythritol	141	100 h	Short-term-test
	Erythritol/Antiox		74 h	Short-term-test
Air	Erythritol/Antiox	141	935 h	Long-term-test
Argon	Erythritol		935 h	Long-term-test

After the samples cooled down to room temperature, they were weighed again. An inert atmosphere was also kept for the samples prepared under argon until they solidified. Finally, each sample was taken out from its vial and processed into a homogeneous mix via manual grinding in a mortar. It is important to note that all samples were already solid when ground to avoid further effects in the melting enthalpy. Keeping an inert atmosphere while the samples were melted was relevant in order to assess if such atmosphere affects the melting enthalpy during the heating.

2.4. Differential Scanning Calorimetry (DSC) Characterization

Latent heat of fusion of the samples after the thermal treatment described in Table 1 was determined via differential scanning calorimetry using a Discovery DSC 2500 manufactured by TA Instruments with an uncertainty of 2%. The DSC cell was purged with 50 mL/min nitrogen while measuring the sample. After the heating procedure described in Section 2.3, sample amounts of about 8–9 mg were placed in Tzero® aluminum pans with aluminum pierced lids. An inert atmosphere was maintained during the whole cycling. Each sample was analyzed in triplicate and the obtained data were averaged. The DSC tests comprised three different cycles. During the first cycle, the sample was heated to 130 °C and then cooled down to 0 °C, both with a heating rate of 10 K/min. Isothermal periods of five minutes were added after reaching both temperature limits (0 °C and 130 °C). The next two cycles were carried out at a heating rate of 1 K/min following the same process of the first cycle. The data were collected from the last heating cycle. Obtained data were analyzed using TRIOS v4.1.1.33073 software.

2.5. Proposed Kinetic Model of Latent Heat Degradation

Thermal degradation of tests performed at different temperatures was evaluated using constant temperature kinetics based on latent heat (L) measured via DSC according to Sagara et al. [26]. A fraction of reaction (α) was defined according to Equation (1) as the change with respect to the initial material property. Thus, L_t is the recorded latent heat after the heat treatment at a certain time, t, while L_0 is the latent heat before the heat treatment.

$$\alpha = 1 - \frac{L_t}{L_0} \tag{1}$$

Kinetics of thermal degradation is usually based on Equation (2):

$$\frac{d\alpha}{dt} = K \cdot f(\alpha) \tag{2}$$

where α is the fraction reacted in time t, and $f(\alpha)$ is the reaction model, which depends on the reaction mechanism according to different physical models [27–29]. Finally, the rate constants (k) for each test were calculated at three different temperatures (121 °C, 131 °C and 141 °C), and the Arrhenius parameters were determined according to Equation (3):

$$K = A \cdot e^{\frac{-E_a}{RT}} \tag{3}$$

2.6. Fourier-Transform Infrared Spectroscopy (FTIR) Analysis

Erythritol thermal stability and changes in its molecular scale, bonds and functional groups after being heated at three different constant temperatures (121 °C, 131 °C and 141 °C) were assessed with a Fourier transformation infrared spectrum analysis (FTIR). FTIR measurements were carried out before and after the thermal treatment in the heating chamber described in Section 2.3. FTIR spectra were recorded on a Spectrum Two™ FT-IR spectrometer from PerkinElmer, in the region 4000–450 cm^{-1}. The obtained data were analyzed with PerkinElmer Spectrum™ 10 software and OriginPro 9.1.

3. Results and Discussion

3.1. Mass Loss

The measured time-dependent mass loss percentages were fitted in a linear regression based on the least squares method. Mass loss rates were represented as the slope of the linear regression, while the uncertainty of each rate was described as the standard error of the given slope. Figure 2a shows the results of all tests.

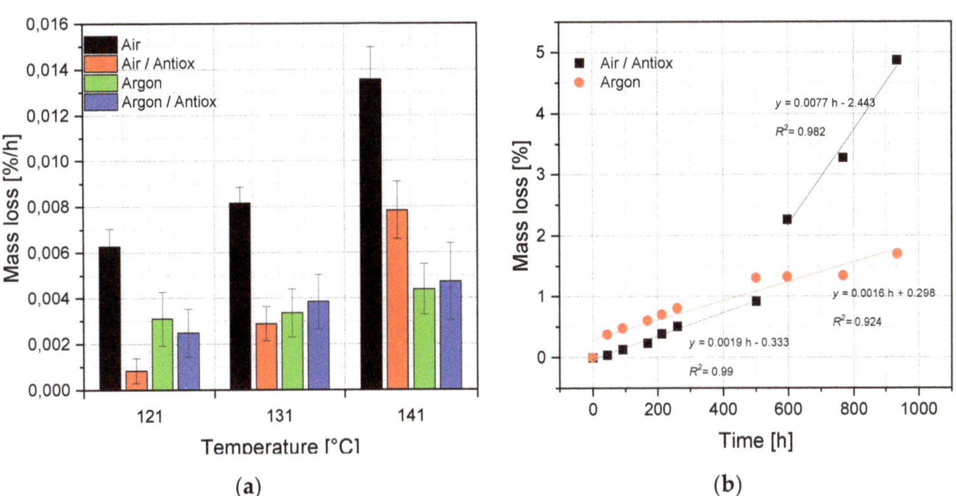

Figure 2. (**a**) Percentage of mass loss per hour at 121 °C, 131 °C and 141 °C for short-term tests. (**b**) Percentage of mass loss at 141 °C for long-term test.

Samples of pure erythritol under air registered the highest mass loss rates along the short-term heating treatment: $0.0063 \pm 0.0008\%/h$, $0.0082 \pm 0.0007\%/h$ and $0.0136 \pm 0.0014\%/h$ at 121 °C, 131 °C and 141 °C, respectively. An upward trend in the percentage of mass decrease regarding heating temperature for both groups of short-term tests under air was registered. The use of antioxidant helped to diminish the temperature effect on the mass content loss of samples under air as seen in Figure 2a. When heated at 141 °C and after 935 h (long-term test) erythritol with antioxidant under air lost nearly 5% of its mass. However, it seems that the antioxidant was consumed during the first 600 h, since after 597 h its mass loss rate tripled the initial one, as shown in Figure 2b.

Regarding samples under argon atmosphere, a mass loss rate increase was also recorded. Temperature, however, had a lighter impact on these results in comparison with the registered values of the samples under air. In addition, adding antioxidant did not have a major impact on the samples under Argon. Vials were weighed before and after filling them with argon, as well as after the heating procedure. As a result of the argon purge, vials were not completely sealed anymore due to the use of cannulas on the lids. Thus, by the end of the experiment it was not possible to assure that the registered mass loss was related only to the sample itself and not to a loss of argon during the heating treatment.

Erythritol mass loss could be associated to oxidation and dehydration reactions. During oxidation, several fragmentation products can be formed (see Section 3.4). Fragmentation produces low molecular weight products with high vapor pressures. Fragmentation products may then evaporate, increasing the mass loss rate. In addition, dehydration of erythritol yields 1,4-erythritane, which can be further dehydrated to form carbonyl compounds also affecting the mass loss ratio. Samples of pure erythritol were more susceptible to oxidization due to the higher oxygen availability in comparison to samples with antioxidants or argon. Dehydration reactions may also have followed. Therefore, mass loss rates of pure erythritol at the three tested temperatures were also the highest. On the other hand, mass loss of both types of samples under argon is more likely related to dehydration reactions of erythritol. Kakiuchi et al. [15] claimed that the degradation of erythritol depends on oxygen concentration. In addition, as it is known, increasing the temperature at which a reaction takes place generally speeds up the rate of reaction. These last two statements can also be confirmed in Figure 2a.

3.2. Latent Heat Degradation

As described in Section 2.4, each sample was analyzed in triplicate and the obtained data were averaged. The standard deviation of each sample was estimated from the calculated average and presented as error bars.

Figure 3a depicts the results of the latent heat degradation for the short-term experiments at 141 °C. Pure erythritol under air suffered the highest latent heat loss over time. A latent heat reduction of 22.7% regarding the initial value was registered. Adding antioxidant helped to improve the thermal stability of samples under air. However, due to the fact that enthalpy is an extensive property, its magnitude is proportional to the amount of the substance that reacts. Therefore, the latent heat value is reduced by the percentage of added antioxidant. The enthalpy of fusion of samples with antioxidant will then be lower than the value of the polyol itself, as it is also shown in Figure 3a. Both experiments under argon atmosphere registered lower latent heat reductions in comparison with erythritol under air. Pure erythritol and erythritol-antioxidant lost 2.8% and 2% of their initial enthalpy values by the end of the test, respectively.

Registered latent heat values were fitted in a linear regression in order to get the degradation rates for all experiments (Figure 3b). An upward trend in the latent heat degradation rate regarding heating temperature was registered for all experiments, with the highest degradation rates for pure erythritol under air. Samples with antioxidant under air showed lower latent heat degradation rates than pure erythritol samples. Latent heat degradation rates of 0.13 ± 0.030 kJ/(kg h), 0.14 ± 0.024 kJ/(kg h) and 0.15 ± 0.021 kJ/(kg h) were registered for samples at 121 °C, 131 °C and 141 °C with antioxidant under air, respectively. This represents about 3 to 4 times less than samples without any additive and under air. Thus, it seems that the temperature rise in combination with antioxidant did not influence the degradation rates of samples under air as much as those of the samples of pure erythritol under air. However, the degradation rate during the long-term test starts to increase from 260 h onwards to a value of 0.19 ± 0.005 kJ/(kg h), which could be due to the gradual consumption of antioxidant during the experiment of 935 h (Figure 4a).

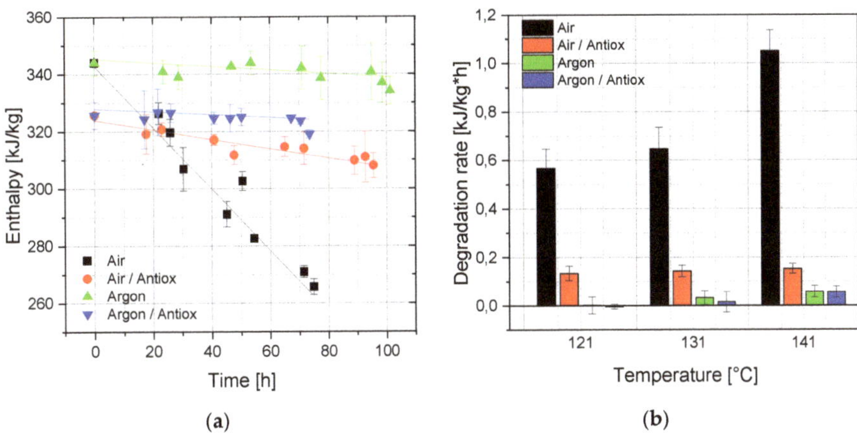

Figure 3. (a) Evolution of latent heat values over time during the short-term experiment at 141 °C. (b) Latent heat degradation rate for short-term tests at three different temperatures.

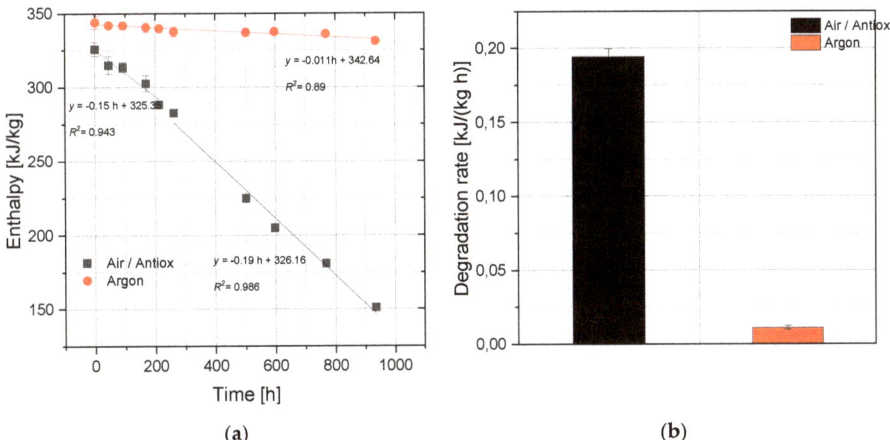

Figure 4. (a) Evolution of latent heat values over time during long-term experiments at 141 °C. (b) Latent heat degradation rate for long-term experiments at 141 °C.

Likewise, Figure 3b shows that lower degradation rates were obtained for both experiments conducted under argon atmosphere, the highest one being 0.06 ± 0.02 kJ/(kg h) for tests at 141 °C. Since the latent heat of fusion recorded from tests under an argon atmosphere decreased less than under air, the slopes of linear fittings were close to zero, resulting in a high standard deviation of the slope. In addition, Figure 4a depicts how the latent heat of fusion of the samples under an argon atmosphere decreased less by the end of the long-term test than those with antioxidant under air, with an average degradation rate of only 0.01 ± 0.001 kJ/(kg h), as shown in Figure 4b. This indicates that an anoxic environment diminishes the erythritol degradation for exposure times up to 935 h.

Latent heat decrease of sugar alcohols after heating has previously been reported [30,31]. It is suggested that the hydrogen bonding structure of erythritol can be weakened and eventually destroyed by temperature increasing, which in turn may cause a decrease in its latent heat [32,33]. Inagaki and Ishida [34] claimed that in the case of sugar alcohols a majority of the released/stored thermal energy originates from the change of electrostatic interaction energy associated with formation/disruption of intermolecular hydrogen bonds in the solid−liquid phase transition. That is, a larger number of intermolecular hydrogen bonds in the solid phase causes a stable solid phase in terms of electrostatic energy. In addition, Matuszek et al. [35] found that latent heat depends on the concentration and strength of hydrogen bonding present in the crystal structure and on the extent of disruption of the crystal phase at elevated temperatures. The same author mentioned that an ideal PCM is one in which many strong hydrogen bonds are present in the crystalline phase and which are readily disrupted on melting. A decrease in latent heat may also be attributed to incomplete crystallization during cooling from the melt. In the case of erythritol, it did not crystallize completely while increasing the temperature of the heating treatment (see Section 3.4). Thus, less energy was required to melt the sample than it would have needed if it had crystallized entirely.

3.3. Kinetic Analysis of Latent Heat Degradation

Kinetics of thermal degradation is usually based on Equation (2). After rearrangement and integration, Equation (2) becomes Equation (4)

$$\int_0^\alpha \frac{d\alpha}{f(\alpha)} = k \int_0^t dt \tag{4}$$

where the left side of the equation is the integrated form of the reaction model $g(\alpha)$, according to Equation (5)

$$\int_0^\alpha \frac{d\alpha}{f(\alpha)} = g(\alpha) \tag{5}$$

Substitution of Equation (5) in Equation (4) leads to Equation (6)

$$g(\alpha) = k \cdot t \tag{6}$$

where $g(\alpha)$ is the integrated reaction model. Selecting the best-fit reaction model in Equation (6) gives a straight line by plotting $g(\alpha)$ versus time t. Nomura et al. [36] showed that a first order reaction, expressed as $-\ln(1 - \alpha)$, is suitable to describe the degradation of the latent heat of sugar alcohols.

As could be seen in Section 3.2, melting enthalpy values obtained from tests under an argon atmosphere remained more stable over time. The 100 h test period was rather short to observe significant latent heat changes. Therefore, the reaction kinetics for tests under an argon atmosphere were not assessed. Figures 5a and 6a depict the data obtained for tests of pure erythritol under air and with antioxidant at the three given temperatures, respectively.

The rate constant (k) was determined from the slope of the straight line. Furthermore, the rate constants (k) were calculated at the three given temperatures and the Arrhenius parameters were determined according to Equation (3). Figures 5b and 6b show the Arrhenius plot for tests of pure erythritol under air and erythritol with antioxidant, respectively.

Reaction rates (k) obtained for erythritol under air increased when the temperature increased. These changes were more evident in samples of pure erythritol. Temperature is a factor that affects reaction rates, i.e., an increase in temperature normally increases the rate of reaction. The temperature dependency of reactions is in turn determined by the activation energy. Activation energy (Ea) values of 41.65 kJ/mol and 5.55 kJ/mol were calculated according to Equation (3) for tests of pure erythritol under air and erythritol with antioxidant, respectively.

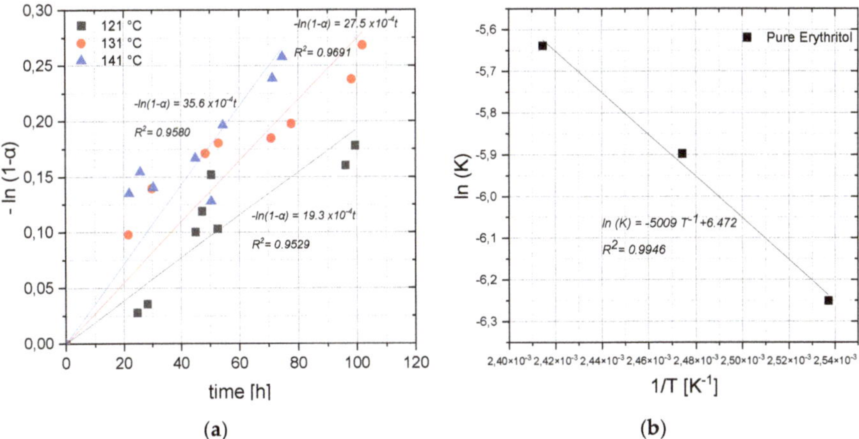

Figure 5. (**a**) Plot of degradation of latent heat $-\ln(1 - \alpha)$ versus time t for pure erythritol under air at three different temperatures. Data points were subjected to a least squares fitting. (**b**) Arrhenius plot for latent heat thermal degradation of pure erythritol under air. Data points were subjected to a least squares fitting.

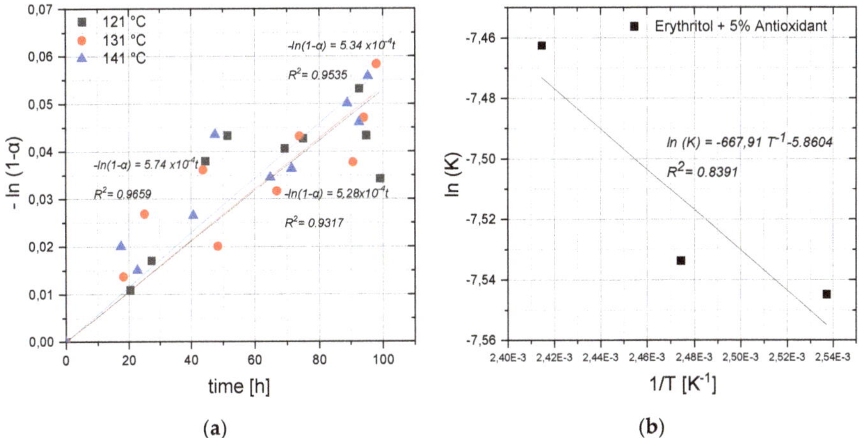

Figure 6. (**a**) Plot of degradation of latent heat $-\ln(1 - \alpha)$ versus time t for erythritol with antioxidant under air at three different temperatures. Data points were subjected to a least squares fitting. (**b**) Arrhenius plot for latent heat thermal degradation of erythritol with antioxidant under air. Data points were subjected to a least squares fitting.

Even though the reaction rates (k) of erythritol with antioxidant tend to increase with increasing temperature as shown in Figure 6a, the effect is not as considerable as it is in pure erythritol samples. Antioxidant was found to diminish the thermal degradation during heating under air and its corresponding samples were less temperature sensitive than those without any added additive. Temperature-sensitive reactions have high activation energies, that is, the higher the activation energy, the faster the reaction is by increasing temperature.

3.4. Chemical Degradation

Samples of pure erythritol under air suffered the most evident physical changes after the thermal treatment. Pure erythritol experienced a significant consistency change, from a loose white powder to a brown sticky paste by the end of the tests. Several samples of pure erythritol under air did not directly solidify after being placed at room temperature.

Erythritol under air became brown with increasing temperature and time under thermal treatment, as can be seen in Figure 7.

Figure 7. Samples of pure erythritol under air for short term at (from left to right) 121 °C, 131 °C and 141 °C.

No exothermic peak was observed in the cooling scan of the sample heated at 141 °C for 78 h, which also did not solidify at room temperature (Figure 8a). This indicates that erythritol became a glass and not a crystal during cooling. The occurrence of a glass transition can be confirmed in the heating scan of the same sample (Figure 8b), in which an exothermic peak can be seen around 6 °C. The latter agrees with the results obtained by other authors [18,37,38]. This exothermic peak corresponds to the cold crystallization of the amorphous phase created in the previous cooling (crystalline phase A). As erythritol was further heated up, a solid/solid transformation took place at around 74 °C, depicted by another small exothermic peak. During solid/solid transformation another crystalline phase similar to the single crystal of meso-erythtritol is formed (crystalline phase B) [38]. Afterward, the final crystalline structure melts at around 120 °C. A solid/solid transformation could take place at three different temperature ranges: about 65 °C, between 15 and 43 °C and about 75 °C. All heating curves of Figure 8b show exothermic peaks around the aforementioned temperature ranges. Each type of solid/solid transformation that takes place is related to the different paths of the erythritol´s crystallization process. The solid/solid transformation has been reported as an entropically driven process, since heat is not involved [18,37]. However, the exothermic peaks that appeared in the heating scans suggest that the solid/solid transformation that occurred involved heat. Similar results were obtained by Nakano et al. [38]. In addition, melting peaks shifted to lower temperatures while increasing the heating treatment time of the sample. Other sugar alcohols have showed similar melting temperature shifting as treatment time or temperature was increased [31,37]. The presence of four hydroxyl groups in erythritol gives rise to a system of intermolecular and intramolecular hydrogen bonds. Molecular structure and properties, such as melting point, are related to hydrogen bonding [33]. The length and strength of hydrogen bonds are sensitive to temperature, as both vary with temperature quadratically [39]. Thus, the melting point could be reduced by weakening the hydrogen bonding when heating the material.

Erythritol with antioxidant under air suffered a slight browning compared with the samples without additive. Samples that were kept longer in the heating chamber suffered the most noticeable changes. On the other hand, pure erythritol under argon barely changed its color along the heating treatment, however, one sample at the highest temperature did not solidify immediately. Sample browning was also not observed in erythritol-antioxidant under argon and all samples solidified at room temperature. Unlike pure erythritol, all erythritol-antioxidant samples under air solidified at room temperature. Burgoa et al. [40] reported that Irganox 1098, a hindered phenolic antioxidant, acted as a nucleating agent in a thermoplastic vulcanizate blend. Zhang et al. [41] claimed that in hindered phenols with a symmetrical structure the formation of hydrogen bonds is high, which leads to agglomeration and crystallization of the system where it is used. Since Irganox 1010 has a high symmetric molecular structure, it may have promoted the crystallization of erythritol. However, further investigation to confirm this latter point is advised.

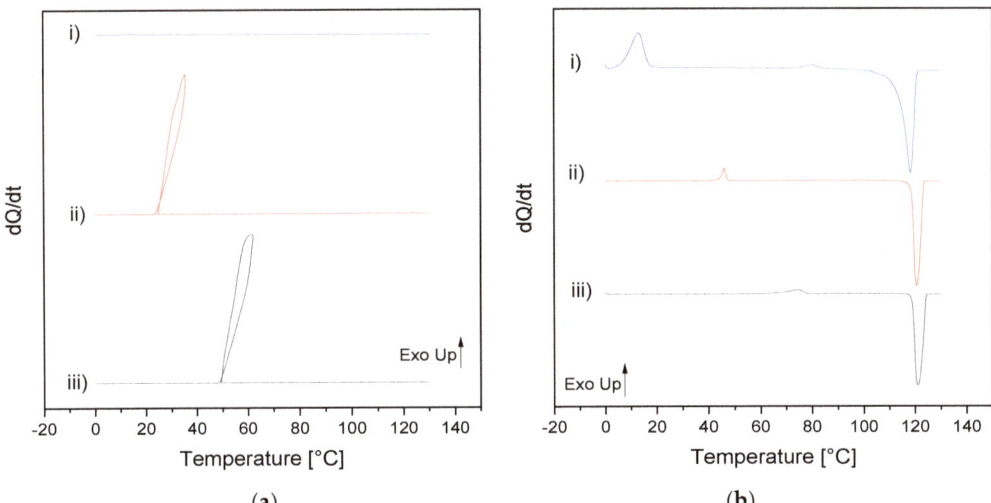

Figure 8. DSC (**a**) cooling and (**b**) heating scans of pure erythritol samples as received (iii), after 50 h (ii) and after 78 h (i) at 141 °C.

It is assumed that erythritol may undergo oxidation and dehydration reactions. In the presence of air and according to the proposed mechanism of oxidation of erythritol based in reactions already reported in the literature [42–45], hydrogen abstraction either from the carbon chain (Pathway A) or from a hydroxyl group (Pathway B) could take place. Pathway A leads to the formation of carbonyl groups by adding oxygenated functional groups to the carbon chain. In pathway B the resulting alkoxy radicals are decomposed via C-C bond scission, generating fragmentation products, as shown in Figure 9. The former case increases vapor pressure by approximately 1 order of magnitude, while the latter may increase volatility by a much larger degree by reducing the carbon number of the product molecules [42].

Irganox 1010 is a primary antioxidant, known as a h-donor and as radical scavenger. In accordance with the mechanism of oxidation of erythritol under air, two different alkyl radicals (R•) can be formed in pathway A, which in turn react with molecular oxygen to produce their corresponding peroxy radicals (ROO•). As mentioned before, oxidation of pathway B also produces alkoxy radicals (RO•). Primary antioxidants terminate oxidation by scavenging R•, ROO• and RO• radicals. In oxygen-rich conditions alkyl radicals react too fast with molecular oxygen and hence their lifetime is shortened enormously, which makes that alkyl radical scavenging become less important [46]. Irganox 1010 (ArOH) converts peroxy and alkoxy radical into inactive products and phenoxy radicals (ArO•). Phenoxy radicals are less reactive; however, they still can react with other ROO•/RO• yielding non-radical products (ROO/ROArO). These non-radical products can rearrange once to regenerate the phenolic antioxidant group, allowing further radicals to be scavenged [46,47]. However, once the antioxidant is fully consumed erythritol will begin to degrade as in the case of the long-term test under air.

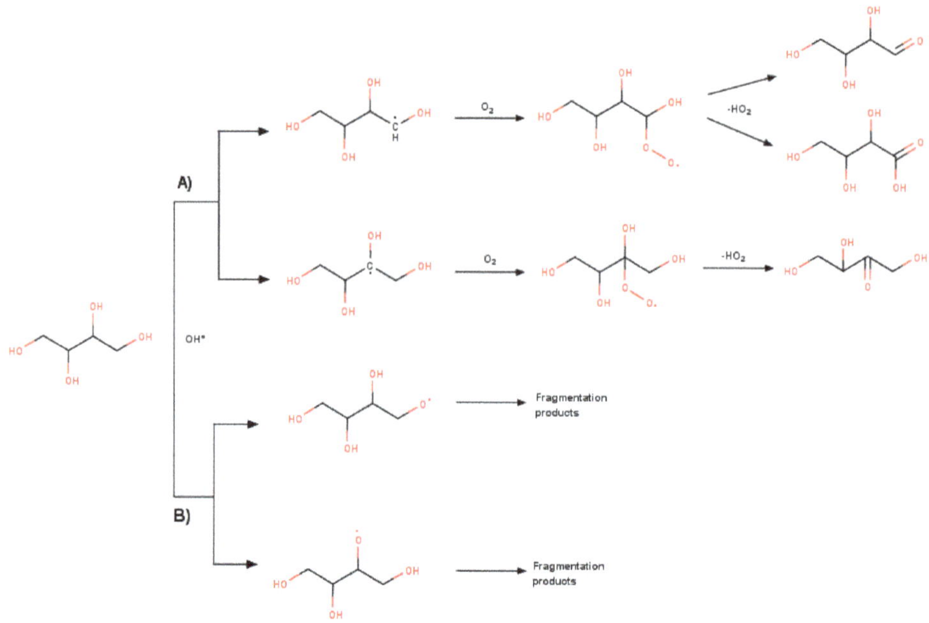

Figure 9. Proposed mechanism of oxidation of erythritol under air. Pathway A: abstraction of hydrogen atom to form alkyl radicals, which in the presence of oxygen are turned into peroxy radicals. Peroxy radicals can undergo the unimolecular HO_2 elimination process yielding carbonyl compounds. Pathway B: abstraction of hydrogen atoms to form alkoxy radicals, which undergo decomposition via C-C bond scission.

It has been reported that erythritol, as polyol, undergoes neither Maillard browning nor caramelization [48–50]. Even though some authors have suggested that the browning of sugar alcohols (D-mannitol) may be related to caramelization reactions [37,51], they have not detected sugar caramelization products, such as hydroximethyl furfural (HMF) and hydroxyacetyl furan (HAF), but α, β-unsaturated carbonyl groups. However, these latter compounds may also be produced during dehydration reactions of polyols. Conjugated ketones, also known as enones, are a type of α, β-unsaturated carbonyl, which are found to be responsible for the color change in various systems. Conjugated ketones may have contributed to the brown coloration of a commercial pentaerythritol tetraester oil [52]. A slight yellowing of the polyvinyl chloride (PVC) was observed due to the formation of conjugated double bonds, which when increasing turned the PVC dark brown to finally black [53]. Color formation was also related to the increase of conjugated double bonds during thermal degradation of poly(L-lactic acid) (PLLA), as its UV-Vis spectra became stronger in intensity while the processing temperature increased [54]. Bayón and Rojas [37] also observed an increase of absorption in the UV-Vis spectra of D-mannitol when kept for longer times at 180 °C under air. In addition, they confirmed the occurrence of α-β unsaturated carbonyl groups, such as 4-methylpent-3-en2-one, which display absorption bands between 240 nm and 320 nm, as a likely indicator of D-mannitol degradation. 4-methylpent-3-en-2-one is a well-known conjugated ketone generated in the aldol condensation of acetone on dehydration [55]. As mentioned before, erythritol became steadily brown with increasing temperature and time of thermal treatment, which is in good agreement with the time/temperature dependency of the formation of α, β-unsaturated carbonyl groups. In addition, Zeitsch [56] claimed that color build up in furfural is due to an increasing number of conjugated double bonds triggered by oxygen. Oxygen triggers discoloration by producing radicals, which attack the double bond of a furfural molecule leading to the formation of an uninterrupted sequence of conjugated double bonds. As already mentioned, pure erythritol under air had the biggest color change after heating treatment, which suggests the formation of conjugated carbonyl systems from the oxidation products. Therefore, it is suggested that erythritol browning is rather related to the formation of conjugated carbonyl compounds when it oxidizes/dehydrates than to the caramelization process.

On the other hand, dehydration of erythritol at high temperatures has been reported [15,52,57]. The experimental method that Kakiuchi et al. [15] described is closer to the one used for this study, as erythritol under air was also heated up to 140 °C and kept there for a period of time. Even though the experimental conditions reported by the above-mentioned authors differ from each other, all of them confirmed 1,4-anhydroerythritol as the first reaction product of the dehydration of erythritol. Thus, not only oxygen concentration, but also temperature and heating time affect the dehydration process of erythritol. Bayón and Rojas [37] mentioned the formation, not only of cyclic ethers, but also of other products through different reaction pathways of thermal dehydration of polyols, giving place to the α, β-unsaturated carbonyl (or enones) groups. Intramolecular dehydration of polyols leads to cyclic ether production via an SN^2 substitution process [58]. Intermolecular dehydration can lead to, among others, carbonyl compounds via a carbocation intermediate (SN^1 mechanism) [59,60]. Ott [61] found the occurrence of carbonyl compounds via enolic intermediate by dehydration at the C-2 position of erythritol. Due to the occurrence of carbonyl compounds, many further possible subsequent reactions can follow, such as aldol additions and condensations, which in turn can give place to the α, β-unsaturated carbonyl groups mentioned also by Bayón and Rojas [37].

Figure 10a shows the FT-IR spectra of samples of pure erythritol under air before and after being heated. As temperature was raised, intermolecular hydrogen bonds weakened, and hence, the hydroxyl bonds strengthened [32]. This caused the –OH absorption band of unheated sugar alcohol (~3225 cm^{-1}) to move to higher wave numbers (up to 3240 cm^{-1}) at higher temperatures. The new peak that appeared at ~1716 cm^{-1} in the thermally treated samples indicates the likely presence of carbonyl groups. In addition, a C=O stretching of a conjugated double bond of a ketone appeared at around 1640 cm^{-1} only for the heated samples [62]. This confirms the formation of conjugated carbonyl groups after heating treatment under air. As proposed, erythritol undergoes oxidation/dehydration reactions when heated under air. Both reactions give rise to carbonyl compounds and conjugated double bonds with similar absorptions bands. Hence, it is hard to confirm which one of the processes was the one that generated the products shown in the FTIR spectra.

Irganox 1010$^{®}$ contains ester linkages [63], which are represented by the absorption peak at ~1730 cm^{-1} of the unheated samples in Figure 10. Heated samples with antioxidant under air (Figure 10c) show weak peak intensities at ~1730 cm^{-1}. This is most likely due to the consumption of antioxidant during thermal treatment, since there was high oxygen availability. In contrast, Figure 10d shows stronger peak intensities at ~1730 cm^{-1} for all heated samples. This could be related to a higher amount of antioxidant remaining in samples, as these were under argon atmosphere, and therefore, there was less oxygen to scavenge.

Unlike samples of pure erythritol under air, the FT-IR spectrum recorded for samples under argon atmosphere (Figure 10b) did not show significant changes either in the shape or in the wave numbers of absorption bands of carbonyl functional groups.

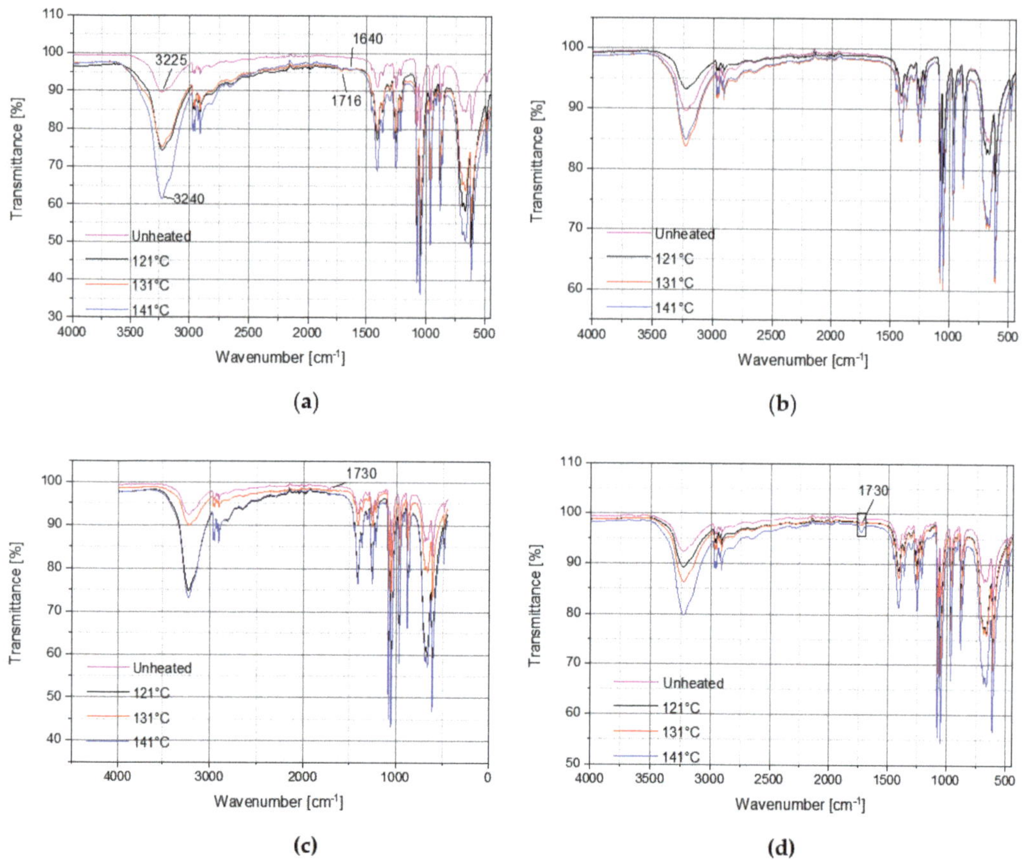

Figure 10. FT-IR spectra of the short-term-tests for (**a**) pure erythritol under air; (**b**) pure erythritol under argon atmosphere; (**c**) erythritol with antioxidant under air and (**d**) erythritol with antioxidant under argon atmosphere.

Lastly, Figure 11a depicts that after around 935 h of thermal treatment, in the long-term tests, two peaks appeared at ~1643 cm^{-1} and ~1725 cm^{-1} in samples of erythritol with antioxidant under air. As previously mentioned, absorption in this region corresponds to carbonyl groups. The C=O stretching of the conjugated ester is no longer visible (~1730 cm^{-1}), since the absorption bands corresponding to the oxidation/caramelization products overlap. However, it is likely that the antioxidant has been depleted during the long-term test, since as shown in Figure 10c, heated samples with antioxidant did not show absorption in the carbonyl groups region, while Figure 11a shows similar absorption bands as Figure 10a, which indicates an oxidation/dehydration process. Heated samples with antioxidant under air show only weaker peaks in the Irganox 1010® absorption region after 100 h of thermal treatment compared to their corresponding unheated sample (Figure 10c). In addition, as mentioned in Section 3.2, during the long-term test and only after 260 h, the degradation rate started to increase, which could be another hint about the antioxidant depletion. On the contrary, Figure 11b shows no chemical changes in pure erythritol under argon atmosphere after the long-term thermal treatment.

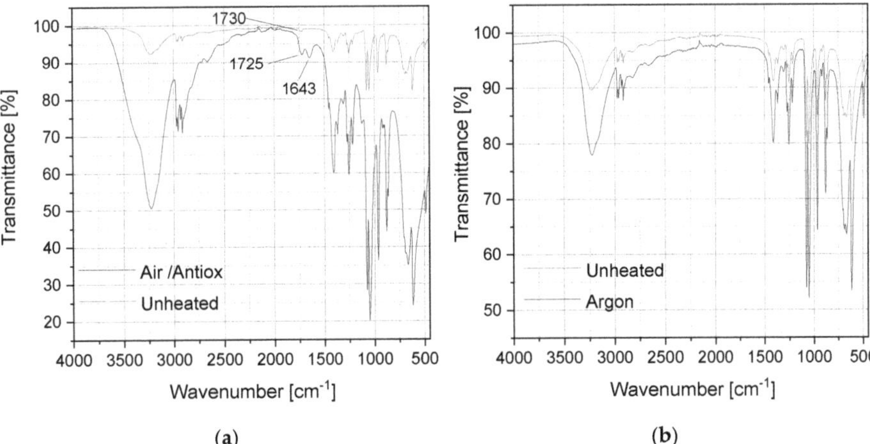

Figure 11. FT-IR spectra of the long-term-tests for (**a**) erythritol with antioxidant under air; (**b**) pure erythritol under argon atmosphere.

4. Conclusions

Sugar alcohols belong to a promising category of organic phase change materials. Erythritol is an attractive sugar alcohol due to its thermal energy storage properties. However, earlier studies indicate that erythritol tends to degrade under thermal cycling. Most of the available studies report the structural changes of erythritol during cycling rather than its thermal stability at constant temperature, which would allow a degradation rate to be determined under conditions closer to real operation. In this study, the effect of thermal treatment above the melting point on the erythritol thermal, chemical and physical properties, as well as its degradation kinetics and the likely thermal stability enhancement by adding antioxidant and using it under inert atmosphere, were addressed.

The latent heat of erythritol under air decreased with the increasing temperature of thermal treatment as its hydrogen bonding structure was likely disrupted at high temperatures. In addition, a decrease in latent heat may also be attributed to incomplete crystallization during cooling from the melt.

Erythritol under air changed its color and some samples remained liquid after thermal treatment. Sample browning upon heating could be related to the formation of conjugated carbonyl compounds during oxidation/dehydration processes. Pure erythritol under air underwent glass transition as no exothermic peak was observed in its DSC thermogram during cooling. Erythritol under argon atmosphere, on the other hand, did solidify and its color barely changed.

Fragmentation products generated during oxidation and carbonyl compounds, among others, generated during dehydration increased the mass loss ratio in samples of pure erythritol under air. Antioxidant helped to reduce this effect.

FT-IR spectra of erythritol under air showed the occurrence of carbonyl groups and conjugated carbonyl compounds after thermal treatment. FT-IR spectra of erythritol under argon did not show new absorption bands, which confirmed its chemical stability after being heated above its melting temperature and up to 141 °C.

Minimal physical changes regarding color and consistency were observed in pure erythritol under argon atmosphere after being thermally treated, thus physical stability could be confirmed for up to 935 h of treatment. Finally, the degradation rate of latent heat was reduced from 0.19 kJ/(kg h) to 0.011 kJ/(kg h) when using an inert atmosphere.

Author Contributions: Conceptualization, M.P.A.L.; methodology, M.P.A.L.; investigation, M.P.A.L.; resources, M.P.A.L.; data curation, M.P.A.L.; writing—original draft preparation, M.P.A.L.; writing—review and editing, H.N. and S.G.; visualization, M.P.A.L.; supervision, H.N.; project administration, S.G.; funding acquisition, S.G. All authors have read and agreed to the published version of the manuscript.

Funding: This research was funded by the German Federal Ministry of Economics and Energy via several research projects, grant number 0325549A.

Institutional Review Board Statement: Not applicable.

Informed Consent Statement: Not applicable.

Data Availability Statement: The data presented in this study are available on request from the corresponding author. The data are not publicly available due to restrictions given by partners.

Acknowledgments: The authors are grateful to the Project Management Jülich for the administrative support.

Conflicts of Interest: The authors declare no conflict of interest.

References

1. Gude, V.G.; Nirmalakhandan, N.; Deng, S.; Maganti, A. Low temperature desalination using solar collectors augmented by thermal energy storage. *Appl. Energy* **2012**, *91*, 466–474. [CrossRef]
2. Tian, Y.; Zhao, C. A review of solar collectors and thermal energy storage in solar thermal applications. *Appl. Energy* **2013**, *104*, 538–553. [CrossRef]
3. N'Tsoukpoe, K.E.; Liu, H.; Le Pierrès, N.; Luo, L. A review on long-term sorption solar energy storage. *Renew. Sustain. Energy Rev.* **2009**, *13*, 2385–2396. [CrossRef]
4. IEA IRENA. Thermal Energy Storage: Technology Brief E17. 2013. Available online: https://www.irena.org/DocumentDownloads/Publications/IRENA-ETSAP20Tech20Brief20E1720Thermal20Energy20Storage.pdf (accessed on 14 September 2018).
5. Garg, H.; Mullick, S.C.; Bhargava, A.K. Sensible heat storage. In *Solar Thermal Energy Storage*; Garg, H.P., Mullick, S.C., Bhargava, A.K., Eds.; Springer: Dordrecht, The Netherlands, 1985; pp. 82–153.
6. Mehling, H.; Cabeza, L.F. *Heat and Cold Storage with PCM: An Up to Date Introduction into Basics and Applications*; Springer: Berlin/Heidelberg, Germany, 2008.
7. Kuznik, F.; David, D.; Johannes, K.; Roux, J.-J. A review on phase change materials integrated in building walls. *Renew. Sustain. Energy Rev.* **2011**, *15*, 379–391. [CrossRef]
8. Ismail, K.; Henríquez, J.R. Thermally effective windows with moving phase change material curtains. *Appl. Therm. Eng.* **2001**, *21*, 1909–1923. [CrossRef]
9. Weinläder, H.; Beck, A.; Fricke, J. PCM-facade-panel for daylighting and room heating. *Sol. Energy* **2005**, *78*, 177–186. [CrossRef]
10. Boussaba, L.; Foufa, A.; Makhlouf, S.; Lefebvre, G.; Royon, L. Elaboration and properties of a composite bio-based PCM for an application in building envelopes. *Constr. Build. Mater.* **2018**, *185*, 156–165. [CrossRef]
11. del Barrio, E.P.; Godin, A.; Duquesne, M.; Daranlot, J.; Jolly, J.; Alshaer, W.; Kouadio, T.; Sommier, A. Characterization of different sugar alcohols as phase change materials for thermal energy storage applications. *Sol. Energy Mater. Sol. Cells* **2017**, *159*, 560–569. [CrossRef]
12. Zhang, H.; Duquesne, M.; Godin, A.; Niedermaier, S.; Del Barrio, E.P.; Nedea, S.V.; Rindt, C.C. Experimental and in silico characterization of xylitol as seasonal heat storage material. *Fluid Phase Equilib.* **2017**, *436*, 55–68. [CrossRef]
13. Gunasekara, S.N.; Pan, R.; Chiu, J.N.; Martin, V. Polyols as phase change materials for surplus thermal energy storage. *Appl. Energy* **2016**, *162*, 1439–1452. [CrossRef]
14. Kaizawa, A.; Maruoka, N.; Kawai, A.; Kamano, H.; Jozuka, T.; Senda, T.; Akiyama, T. Thermophysical and heat transfer properties of phase change material candidate for waste heat transportation system. *Heat Mass Transf.* **2008**, *44*, 763–769. [CrossRef]
15. Kakiuchi, H.; Yamazaki, M.; Yabe, M.; Chihara, S.; Terunuma, Y.; Sakata, Y.; Usami, T. A study of erythritol as phase change material. In Proceedings of the 2nd Workshop IEA Annex 10, Phase Change Materials and Chemical Reactions for Thermal Energy Storage, Sofia, Bulgaria, 11–13 April 1998.
16. Nomura, T.; Tsubota, M.; Oya, T.; Okinaka, N.; Akiyama, T. Heat release performance of direct-contact heat exchanger with erythritol as phase change material. *Appl. Therm. Eng.* **2013**, *61*, 28–35. [CrossRef]
17. Zhang, H.; van Wissen, R.M.J.; Nedea, S.V.; Rindt, C.C.M. Characterization of sugar alcohols as seasonal heat storage media—experimental and theoretical investigations. In Proceedings of the Advances in Thermal Energy Storage, EUROTHERM 99, Lleida, Spain, 28–30 May 2014.
18. Jesus, A.J.L.; Nunes, S.C.; Silva, M.R.; Beja, A.M.; Redinha, J. Erythritol: Crystal growth from the melt. *Int. J. Pharm.* **2010**, *388*, 129–135. [CrossRef] [PubMed]
19. Diarce, G.; Gandarias, I.; Campos-Celador, Á.; García-Romero, A.; Griesser, U. Eutectic mixtures of sugar alcohols for thermal energy storage in the 50–90 °C temperature range. *Sol. Energy Mater. Sol. Cells* **2015**, *134*, 215–226. [CrossRef]
20. Guo, S.; Liu, Q.; Zhao, J.; Jin, G.; Wang, X.; Lang, Z.; He, W.; Gong, Z. Evaluation and comparison of erythritol-based composites with addition of expanded graphite and carbon nanotubes. *Appl. Energy* **2017**, *205*, 703–709. [CrossRef]

21. Agyenim, F.; Eames, P.; Smyth, M. Experimental study on the melting and solidification behaviour of a medium temperature phase change storage material (Erythritol) system augmented with fins to power a LiBr/H$_2$O absorption cooling system. *Renew. Energy* **2011**, *36*, 108–117. [CrossRef]

22. Kaizawa, A.; Kamano, H.; Kawai, A.; Jozuka, T.; Senda, T.; Maruoka, N.; Akiyama, T. Thermal and flow behaviors in heat transportation container using phase change material. *Energy Convers. Manag.* **2008**, *49*, 698–706. [CrossRef]

23. BASF. Industrial Coatings: Technical Data Sheet. Irganox 1010. 2015. Available online: https://worldaccount.basf.com/wa/NAFTA~{}en_US/Catalog/Additives/info/BASF/PRD/30546637 (accessed on 18 March 2019).

24. Zaharescu, T.; Giurginca, M.; Jipa, S. Radiochemical oxidation of ethylene–propylene elastomers in the presence of some phenolic antioxidants. *Polym. Degrad. Stab.* **1999**, *63*, 245–251. [CrossRef]

25. Gensler, R.; Plummer, C.; Kausch, H.-H.; Krämer, E.; Pauquet, J.-R.; Zweifel, H. Thermo-oxidative degradation of isotactic polypropylene at high temperatures: Phenolic antioxidants versus HAS. *Polym. Degrad. Stab.* **2000**, *67*, 195–208. [CrossRef]

26. Sagara, A.; Nomura, T.; Tsubota, M.; Okinaka, N.; Akiyama, T. Improvement in thermal endurance of D-mannitol as phase-change material by impregnation into nanosized pores. *Mater. Chem. Phys.* **2014**, *146*, 253–260. [CrossRef]

27. Vyazovkin, S.; Sbirrazzuoli, N. Kinetic methods to study isothermal and nonisothermal epoxy-anhydride cure. *Macromol. Chem. Phys.* **1999**, *200*, 2294–2303. [CrossRef]

28. Vyazovkin, S.; Wight, C.A. Model-free and model-fitting approaches to kinetic analysis of isothermal and nonisothermal data. *Thermochim. Acta* **1999**, *340*, 53–68. [CrossRef]

29. Halikia, I.; Zoumpoulakis, L.; Christodoulou, E.; Prattis, D. Kinetic study of the thermal decomposition of calcium. *Eur. J. Miner. Process. Environ. Prot.* **2001**, *1*, 89–102.

30. Neumann, H.; Niedermaier, S.; Gschwander, S.; Schossig, P. Cycling stability of d-mannitol when used as phase change material for thermal storage applications. *Thermochim. Acta* **2018**, *660*, 134–143. [CrossRef]

31. Solé, A.; Neumann, H.; Niedermaier, S.; Martorell, I.; Schossig, P.; Cabeza, L.F. Stability of sugar alcohols as PCM for thermal energy storage. *Sol. Energy Mater. Sol. Cells* **2014**, *126*, 125–134. [CrossRef]

32. Feng, H.; Liu, X.; He, S.; Wu, K.; Zhang, J. Studies on solid-solid phase transitions of polyols by infrared spectroscopy. *Thermochim. Acta* **2000**, *348*, 175–179. [CrossRef]

33. Wu, N.; Li, X.; Liu, S.; Zhang, M.; Ouyang, S. Effect of hydrogen bonding on the surface tension properties of binary mixture (acetone-water) by Raman spectroscopy. *Appl. Sci.* **2019**, *9*, 1235. [CrossRef]

34. Inagaki, T.; Ishida, T. Computational design of non-natural sugar alcohols to increase thermal storage density: Beyond existing organic phase change materials. *J. Am. Chem. Soc.* **2016**, *138*, 11810–11819. [CrossRef]

35. Matuszek, K.; Vijayaraghavan, R.; Kar, M.; Macfarlane, D.R. Role of hydrogen bonding in phase change materials. *Cryst. Growth Des.* **2019**, *20*, 1285–1291. [CrossRef]

36. Nomura, T.; Zhu, C.; Sagara, A.; Okinaka, N.; Akiyama, T. Estimation of thermal endurance of multicomponent sugar alcohols as phase change materials. *Appl. Therm. Eng.* **2015**, *75*, 481–486. [CrossRef]

37. Bayón, R.; Rojas, E. Feasibility study of D-mannitol as phase change material for thermal storage. *AIMS Energy* **2017**, *5*, 404–424. [CrossRef]

38. Nakano, K.; Masuda, Y.; Daiguji, H. Crystallization and melting behavior of erythritol in and around two-dimensional hexagonal mesoporous silica. *J. Phys. Chem. C* **2015**, *119*, 4769–4777. [CrossRef]

39. Dougherty, R.C. Temperature and pressure dependence of hydrogen bond strength: A perturbation molecular orbital approach. *J. Chem. Phys.* **1998**, *109*, 7372–7378. [CrossRef]

40. Burgoa, A.; Hernandez, R.; Vilas, J.L. New ways to improve the damping properties in high-performance thermoplastic vulcanizates. *Polym. Int.* **2020**, *69*, 467–475. [CrossRef]

41. Zhang, L.; Chen, D.; Fan, X.; Cai, Z.; Zhu, M. Effect of hindered phenol crystallization on properties of organic hybrid damping materials. *Materials* **2019**, *12*, 1008. [CrossRef] [PubMed]

42. Kessler, S.H.; Smith, J.D.; Che, D.L.; Worsnop, D.R.; Wilson, K.R.; Kroll, J.H. Chemical sinks of organic aerosol: Kinetics and products of the heterogeneous oxidation of erythritol and levoglucosan. *Environ. Sci. Technol.* **2010**, *44*, 7005–7010. [CrossRef] [PubMed]

43. George, I.J.; Abbatt, J.P.D. Heterogeneous oxidation of atmospheric aerosol particles by gas-phase radicals. *Nat. Chem.* **2010**, *2*, 713–722. [CrossRef] [PubMed]

44. Kroll, J.H.; Lim, C.; Kessler, S.H.; Wilson, K.R. Heterogeneous oxidation of atmospheric organic aerosol: Kinetics of changes to the amount and oxidation state of particle-phase organic carbon. *J. Phys. Chem. A* **2015**, *119*, 10767–10783. [CrossRef]

45. den Hartog, G.J.; Boots, A.W.; Adam-Perrot, A.; Brouns, F.; Verkooijen, I.W.; Weseler, A.R.; Haenen, G.R.; Bast, A. Erythritol is a sweet antioxidant. *Nutrients* **2010**, *26*, 449–458. [CrossRef] [PubMed]

46. Gijsman, P. Polymer stabilization. In *Handbook of Environmental Degradation of Materials*, 3rd ed.; Kutz, M., Ed.; William Andrew: Oxford, UK, 2018; pp. 369–395.

47. Frankel, E.N. (Ed.) Chapter 9—Antioxidants. In *Lipid Oxidation*, 2nd ed.; Oily Press Lipid Library Series; Woodhead Publishing: Cambridge, UK, 2012; pp. 209–258.

48. Ghosh, S.; Sudha, M.L. A review on polyols: New frontiers for health-based bakery products. *Int. J. Food Sci. Nutr.* **2011**, *63*, 372–379. [CrossRef]

49. Grembecka, M. Sugar alcohols as sugar substitutes in food industry. In *Sweeteners: Pharmacology, Biotechnology, and Applications*; Mérillon, J.-M., Ramawat, K.G., Eds.; Springer International Publishing: Cham, Switzerland, 2018; pp. 547–573.
50. Hartel, R.W.; von Elbe, J.H.; Hofberger, R. Chemistry of bulk sweeteners. In *Confectionery Science and Technology*; Springer Science and Business Media LLC: Berlin, Germany, 2017; pp. 3–37.
51. Rodríguez-García, M.-M.; Bayón, R.; Rojas, E. Stability of D-mannitol upon melting/freezing cycles under controlled inert atmosphere. *Energy Procedia* **2016**, *91*, 218–225. [CrossRef]
52. Karis, T.E.; Miller, J.L.; Hunziker, H.E.; de Vries, M.S.; Hopper, D.A.; Nagaraj, H.S. Oxidation chemistry of a pentaerythritol tetraester oil. *Tribol. Trans.* **1999**, *42*, 431–442. [CrossRef]
53. Schiller, M. PVC Stabilizers. In *PVC Additives: Performance, Chemistry, Developments and Sustainability*; Hanser: Munich, Germany, 2015; pp. 1–114.
54. Wang, Y.; Steinhoff, B.; Brinkmann, C.; Alig, I. In-line monitoring of the thermal degradation of poly(l-lactic acid) during melt extrusion by UV–vis spectroscopy. *Polymer* **2008**, *49*, 1257–1265. [CrossRef]
55. Pauwels, D.; Hereijgers, J.; Verhulst, K.; de Wael, K.; Breugelmans, T. Investigation of the electrosynthetic pathway of the aldol condensation of acetone. *Chem. Eng. J.* **2016**, *289*, 554–561. [CrossRef]
56. Zeitsch, K.J. (Ed.) The discoloration of furfural. In *The Chemistry and Technology of Furfural and Its Many By-Products*; Elsevier: Amsterdam, The Netherlands, 2000; pp. 28–33.
57. Ott, L.; Lehr, V.; Urfels, S.; Bicker, M.; Vogel, H. Influence of salts on the dehydration of several biomass-derived polyols in sub- and supercritical water. *J. Supercrit. Fluids* **2006**, *38*, 80–93. [CrossRef]
58. Yamaguchi, A.; Muramatsu, N.; Mimura, N.; Shirai, M.; Sato, O. Intramolecular dehydration of biomass-derived sugar alcohols in high-temperature water. *Phys. Chem. Chem. Phys.* **2016**, *19*, 2714–2722. [CrossRef]
59. Vilcocq, L.; Cabiac, A.; Especel, C.; Guillon, E.; Duprez, D. Transformation of sorbitol to biofuels by heterogeneous catalysis: Chemical and industrial considerations. *Oil Gas Sci. Technol. Rev. IFP Energ. Nouv.* **2013**, *68*, 841–860. [CrossRef]
60. Kurszewska, M.; Skorupa, E.; Kasprzykowska, R.; Sowiński, P.; Wiśniewski, A. The solvent-free thermal dehydration of tetritols on zeolites. *Carbohydr. Res.* **2000**, *326*, 241–249. [CrossRef]
61. Ott, L. Stoffliche Nutzung von Biomasse mit Hilfe von nah- und überkritischen Wasser: Homogenkatalysierte Dehydratisierung von Polyoen zu Aldehyden. Ph.D. Thesis, Technische Universität Darmstadt, Darmstadt, Germany, 2005.
62. Coates, J. Interpretation of infrared spectra, a practical approach. In *Encyclopedia of Analytical Chemistry: Applications, Theory and Instrumentation*; Meyers, R.A., Ed.; John Wiley & Sons Ltd.: Chichester, UK, 2000.
63. Ratnam, C.T.; Nasir, M.; Baharin, A.; Zaman, K. Electron-beam irradiation of poly(vinyl chloride)/epoxidized natural rubber blend in the presence of Irganox 1010. *Polym. Degrad. Stab.* **2001**, *72*, 147–155. [CrossRef]

Article

Selection of a PCM for a Vehicle's Rooftop by Multicriteria Decision Methods and Simulation

Juan Francisco Nicolalde [1,*], Mario Cabrera [1], Javier Martínez-Gómez [1,2], Rodger Benjamín Salazar [3] and Evelyn Reyes [1]

[1] Facultad de Ingeniería y Ciencias Aplicadas, Universidad Internacional SEK, Quito 170302, Ecuador; mcabrera.mec@uisek.edu.ec (M.C.); javier.martinez@uisek.edu.ec (J.M.-G.); epreyes.mee@uisek.edu.ec (E.R.)
[2] Instituto de Investigación Geológico y Energético (IIGE), Quito 170518, Ecuador
[3] Facultad de Ciencias de la Ingeniería, Universidad Técnica Estatal de Quevedo, Quevedo 120301, Ecuador; rsalazarl@uteq.edu.ec
* Correspondence: juan.nicolalde@uisek.edu.ec

Featured Application: Energy efficiency and thermal comfort on vehicles.

Abstract: The automotive industry is one of the most contaminant; for this reason, solutions in efficient matter has been proposed over the years. This research contributes to this subject by evaluating the thermal comfort in the internal air of a vehicle by using a 20 mm layer of a phase-change material attached to the rooftop interior of a car. The phase-change material selection is based on a list of other materials proposed in previous research and chosen by multicriteria decision methods. In this sense, the material savENRG PCM-HS22P proved to be the best. Moreover, a simulation using the finite elements method showed how the PCM reduced the temperature of the air by 9 °C when heating and by 4 °C when the temperature drops. To conclude, the multicriteria selection methods chose the best material to absorb energy during the charging process and released it during the discharging event in this automotive application.

Keywords: phase-change material; multicriteria decision; finite element; automotive; energy storage

Citation: Nicolalde, J.F.; Cabrera, M.; Martínez-Gómez, J.; Salazar, R.B.; Reyes, E. Selection of a PCM for a Vehicle's Rooftop by Multicriteria Decision Methods and Simulation. *Appl. Sci.* **2021**, *11*, 6359. https://doi.org/10.3390/app11146359

Academic Editor: Ioannis Kartsonakis

Received: 27 May 2021
Accepted: 24 June 2021
Published: 9 July 2021

Publisher's Note: MDPI stays neutral with regard to jurisdictional claims in published maps and institutional affiliations.

1. Introduction

The utilization of air conditioning and heating systems to control passengers' comfort in the automotive industry impacts the fuel consumption and economy, which makes it necessary to improve the fuel efficiency [1–3], where the internal temperature of the vehicle's air, as a measurable factor of discomfort, should be controlled between 23 °C and 28 °C [4]. In order to handle this, the development of energy-efficient solutions have been implemented, one of these being Thermal Energy Storage (TES), which is the most efficient in using the available heat resources [5,6]. In this sense, Latent Heat Storage (LHS) is responsible for the accumulation of energy. In phase-change materials (PCM), this phenomena takes place at a molecular level and triggers the transition between phases [7], producing an endothermic reaction when melting and an exothermic reaction when changing from liquid to solid [8]. In this sense, the performance of the PCM depends on the climate and the amount of PCM [9]. Moreover, when talking about LHS, PCMs are classified in organic PCMs, inorganic PCMs and eutectic mixtures; these last ones are mixes of two or more composites that melt and solidify together [10].

The utilization of PCMs has been widely studied for the properties presented during the phase change; in this sense, naming the research of Bakan et al., it has been stablished the importance of the crystallization during the phase change and the requirement to study the crystal growth in different temperature ranges. In this way, the crystal growth of the PCM $Ge_2Sb_2Te_2$, in a range of temperatures between 300 K and 870 K, has been studied [11], allowing to develop a PCM with promising nanophotonic applications that has to be investigated [12]. Furthermore, the thermal applications of PCMs were investigated to reduce the indoor temperatures and reduce the internal air temperature to reach a maximum [13] by storing solar energy [14]. On the other hand, in the automotive industry, PCMs have been used as thermo stabilizers for batteries of electric and hybrid vehicles, by storing the overheat since natural and forced convection are not as efficient [6]. The size of the radiator and the cooling fan also can be reduced by using PCMs and help the cold ignite as well as recover the latent heat [15,16].

PCMs have a wide variety and different properties that are needed for different applications [17]; in this sense, multicriteria decision methods (MCDM) have been used in the selection of PCMs over several areas with great results. The methods used in this research are the Analytic Hierarchy Process [18], the Technique for Order Preference by Similarity to Ideal Solution (TOPSIS) [19], the VIKOR method [20] and the Complex Proportional Assessment Methods (COPRAS) [21]. Moreover, the selection by MCDM was validated by simulations of the superplastic forming for a vehicle's components [22], as well as the selection by MCDM and simulation to enhance the PCM by nanoparticles [23]. In this sense, finite elements analysis has been useful in automotive design, such as in disk brakes and its thermo-mechanical behavior [24].

The building industry has had a leading role in the research of simulated PCMs selected by MCDM, but the automotive sector does not much apply these tools for this benefit. With this previous knowledge, this research aims to select the best PCM by MCDM means based on a bibliographical research, to be applied in a 20 mm layer of PCM that will store energy in the charging process and release it in the discharging event to control the internal air temperature, and where this phenomenon will be simulated by computer-aided engineering.

2. Materials and Methods

2.1. Materials Determmination

Rastogi et al. did a research where 35 PCMs passed through a MCDM selection, where the candidates were considered for heating, ventilation and air-conditioning application in buildings; also in this investigation, the author takes in consideration the Figures of Merit (FOMs), for the performance of the heat extraction per unit volume (FOM1) and the response time of the material (FOM2). Thus, these criteria will be used as the base, where 15 with the best specific heat capacity will be considered for the MCDM proposed for automotive rooftop applications. However, in this selection, some materials from the same family have the same specific heat capacity; for these materials, the one with the best heat extracted per unit volume (FOM1) calculated is considered. In this way, the materials PlusICE PCM A22, PlusICE PCM A23, PlusICE PCM S19, PlusICE PCM S21, PlusICE PCM S25, Rubitherm GmbH PCM SP21E2, Rubitherm GmbH PCM SP25E2, Rubitherm GmbH PCM RT21, Rubitherm GmbH PCM RT24, Rubitherm GmbH PCM RT25, Rubitherm GmbH PCM RT27, Rubitherm GmbH PCM RT21HC and Rubitherm GmbH PCM RT22HC are rejected and, in the case of the materials Rubitherm GmbH PCM SP24E and Rubitherm GmbH PCM SP26E that have the same values, the series SP24E stands for its wide range of phase-change temperatures [25]. In this sense, Table 1 displays the materials considered with its thermal properties. Furthermore, it is important to point out that this investigation will use the lowest thermal phase-change temperature and characteristics in the liquid state; also, an M index was added to the materials as a label to be used in the MCDM to be used instead of the full name. On the other hand, Table 2 shows the results of the calculations made for the figures of merit [25].

2.2. Analytic Hierarchy Process (AHP)

For this automotive application, the following characteristics are searched for:

1. Phase change in an environment temperature;
2. Good density for a low volume change when changing phases;
3. Low fusion heat for a quick and efficient phase change;
4. High latent heat for efficient thermal storage;
5. Good thermal conductivity to transmit the thermal energy.

Table 1. List of phase-change materials and their thermo-physical properties.

Compound (M)		Phase-Change Temp (°C)	Density (kg/m³)	Heat of Fusion (kJ/kg)	Specific Heat Capacity (kJ/kgK)	Thermal Conductivity (W/mK)
RUBITHERM GmbH, PCM SP 24 E	M1	24–35	1500	190	2	0.6
PlusICE PCM, S23	M2	23	1530	175	2.2	0.54
savENRG PCM-HS24P	M3	24	1820	185	2.26	0.5–1.09
PUR-PCM, BASF Polyurethanes GmbH	M4	22	970	365	2	0.19
PlusICE PCM, S17	M5	17	1525	160	1.9	0.43
CoolZONE23, Armstrong	M6	21–22	770	342	2	0.2
savENRG PCM-HS22P	M7	23	1540	185	3.05	0.5–1.09
ThermalCORE 23 C/ 73 F, USA	M8	22–24	770	342	2.2	0.2
Weber.murclima 23, St. Gobain-weber	M9	22–24	950	170	2.32	0.38
RUBITHERM GmbH, PCM RT 25 HC	M10	22–26	880	230	2	0.2
PlusICE PCM, PCM, A25H	M11	25	810	226	2.15	0.18
PlusICE PCM, A22H	M12	22	820	216	2.85	0.18
PlusICE PCM, A25	M13	25	785	150	2.26	0.18
PlusICE PCM, A24	M14	24	790	145	2.22	0.18
PCM-Akustikputz 23, SchreffGmbh& Co.	M15	21–22	400	196	1.7	0.08

Table 2. FOMs results.

Compound (M)		$FOM1*10^6\rho*L$	$FOM2*10^{-6}k/\rho*C_p$
RUBITHERM GmbH, PCM SP 24 E	M1	285	0.2
PlusICE PCM, S23	M2	267.75	0.16042
savENRG PCM-HS24P	M3	336.7	0.12161
PUR-PCM, BASF Polyurethanes GmbH	M4	354.05	0.09793
PlusICE PCM, S17	M5	244	0.14840
CoolZONE23, Armstrong	M6	263.34	0.12987013
savENRG PCM-HS22P	M7	284.9	0.10631
ThermalCORE 23 C/ 73 F, USA	M8	263.34	0.11806
Weber.murclima 23, St. Gobain-weber	M9	161.5	0.17241
RUBITHERM GmbH, PCM RT 25 HC	M10	202.4	0.11363
PlusICE PCM, PCM, A25H	M11	183.06	0.10335
PlusICE PCM, A22H	M12	177.12	0.07702
PlusICE PCM, A25	M13	117.75	0.101459
PlusICE PCM, A24	M14	114.55	0.10263428
PCM-Akustikputz 23, SchreffGmbh& Co.	M15	78.4	0.11764

These requirements are fulfilled by the candidates and the AHP method allows to weight them; this process is developed in the research of Odu. G.O. [26]. Moreover, the method requires a label for the criteria, which is displayed as follows:

- Phase-change Temp = T1;
- Density = T2;
- Heat of fusion= T3;
- Specific heat capacity = T4;
- Thermal conductivity = T5;
- $FOM1 * 10^6\rho * L$ = T6;
- $FOM2 * 10^{-6}k/\rho * C_p$ = T7.

2.3. Method VIKOR

Following the method used by Shekhovtsov and Salabun [27], the VIKOR method solves selection problems looking for a ranking where the alternative and criteria that will be used is the solution closest to the ideal [28]. This method requires that the criteria described before have the following considerations:

- T1 Higher = Better;

- T2 Lower = Better;
- T3 Higher = Better;
- T4 Higher = Better;
- T5 Higher = Better;
- T6 Higher = Better;
- T7 Lower = Better.

With these in mind, the VIKOR method will be performed as the named previous research.

2.4. TOPSIS Method

The TOPSIS method takes and classifies a finite number of alternatives by similarity to the ideal solutions, looking for the alternatives to have the shortest distance to the ideal positive solution and the farthest from the negative [28]. The steps proposed by Shekhovtsov and Salabun [27] are followed and the cost and profit criteria are the same that were determined by the VIKOR method.

2.5. COPRAS-G Method

The ranking of the alternatives in the COPRAS method considers the utility degree of the different options by using grey numbers that comes from the grey theory for insufficient information [28]. This method is followed as in the research of Mousavi-nasab [29] and Sotoudeh-anvai [30], where the beneficial and non-beneficial criteria are defined as follows:

- T1 = Beneficial;
- T2 = Non-Beneficial;
- T3 = Beneficial;
- T4 = Beneficial;
- T5 = Beneficial;
- T6 = Beneficial;
- T7 = Non-Beneficial.

2.6. Spearman's Correlation Coefficient

The different methods obtains different results; to measure the relation between these non-linear results, Spearman's correlation was used. In this sense, this technique quantifies the strength between the variables; if there are no duplicated data, the perfect correlation is +1 or −1. This was calculated following the research of Beltrán and Martínez-Gómez [28].

2.7. Simulation

The simulation of the system takes into consideration the three solids that make up the rooftop of the vehicle, which is made by steel bake hardening, YS260, and cold rolling. This material is described as used for automotive applications, such as roofs, in the software CES-Granta Edupack [31]. The next body represents a layer of material that corresponds to the 20 mm PCM made of the selected material by the MCDM and finally air fills the space; the proposed geometry is displayed in Figure 1. On the other hand, the simulation took two events, the first when the roof is heating up and the PCM is storing energy and the second where the roof is cooling down and the PCM releases the stored energy.

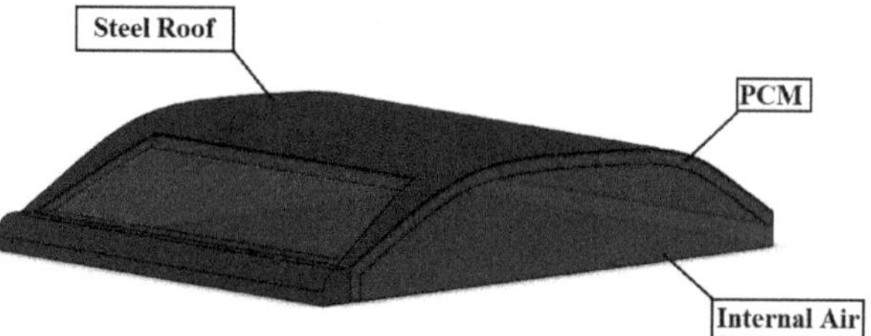

Figure 1. The CAD model.

2.8. Boundary Conditions

The simulation of the charging events comes with the parameters investigated by Dadour et al., who found that the temperature of a parked vehicle's black rooftop can start at 12 °C and rise to 45 °C, provoking a charge event [32]. On the other hand, after reaching this heating peak, a cooling process begins and the temperature reduces to 12 °C in 15 h. Where a discharging event takes place, these events are the ones that will be simulated [32]. Moreover, a convection process with a coefficient for free gases is found in the range of $2\frac{W}{m^2K}$ and $25\frac{W}{m^2K}$ [33]; for this reason, a middle value of $12\frac{W}{m^2K}$ will be used to simulate this phenomena. It is important to point out that the initial temperature of the air is the one reached in the charging event; the boundary conditions of these simulations are displayed in Tables 3 and 4 with the mesh data of the simulation.

Table 3. Boundary conditions for charge.

Element	Parameter
Roof Initial Temperature	12 °C
Roof Final Temperature	45 °C
Temperature Time Lapse	5 h
PCM Initial Temperature	12 °C
Internal Air Initial Temperature	12 °C
Environmental Temperature	24 °C
Convection coefficient	12 W/m^2 * K
Software	Solidworks 2020
Mesher	Blended curvature-based mesh
Mesh Quality	High
Jacobian Points	4
Max element size	22 mm

Table 4. Boundary conditions for discharge.

Element	Parameter
Roof Initial Temperature	45 °C
Roof Final Temperature	12 °C
Temperature Time Lapse	15 h
PCM Initial Temperature	40 °C
Internal Air Initial Temperature	37 °C
Environmental Temperature	24 °C
Convection coefficient	12 W/m^2 * K
Software	Solidworks 2020
Mesher	Blended curvature-based mesh
Mesh Quality	High
Jacobian Points	4
Max element size	22 mm

3. Results and Discussion

3.1. AHP Results

The different processes are shown in the following tables, where Table 5 presents the pair-wise assessment, Table 6 displays the normalization of these and Table 7 exhibits the weighted results of the AHP method, where specific heat capacity has a more significant importance in this application since this property absorbs heat and releases it.

Table 5. Comparison matrix of the criteria.

Criteria	Phase-Change Temp (°C)	Density (kg/m^3)	Heat of Fusion (kJ/kg)	Specific Heat Capacity (kJ/kgK)	Thermal Conductivity (W/mK)	FOM1	FOM2
Phase-change Temp (°C)	1	3	3	0.33	0.33	0.33	0.33
Density (kg/m^3)	0.33	1	0.2	0.20	0.33	0.33	0.33
Heat of fusion (kJ/kg)	0.33	5	1	0.33	0.33	1	1
Specific heat capacity (kJ/kgK)	3	5	3	1	3	3	3
Thermal conductivity (W/mK)	3	3	3	0	1	1	3
FOM1	3	3	1	0	1.00	1	1
FOM2	3	3	1	0	0.33	1	1
Summatory	13.67	23	12.20	2.87	6.33	7.67	9.67

Table 6. Normalized matrix of the criteria.

Phase-Change Temp (°C)	Density (kg/m^3)	Heat of Fusion (kJ/kg)	Specific Heat Capacity (kJ/kgK)	Thermal Conductivity (W/mK)	FOM1	FOM2
0.0732	0.1304	0.2459	0.12	0.0526	0.04	0.03
0.0244	0.0435	0.0164	0.07	0.0526	0.04	0.03
0.0244	0.2174	0.0820	0.12	0.0526	0.13	0.10
0.2195	0.2174	0.2459	0.3488	0.4737	0.39	0.31
0.2195	0.1304	0.2459	0.12	0.1579	0.13	0.31
0.2195	0.1304	0.0820	0.1163	0.1579	0.13	0.10
0.2195	0.1304	0.0820	0.1163	0.0526	0.13	0.10

Table 7. Weighted criteria.

Criteria	Phase-Change Temp (°C)	Density (kg/m^3)	Heat of Fusion (kJ/kg)	Specific Heat Capacity (kJ/kgK)	Thermal Conductivity (W/mK)	FOM1	FOM2
Compound	T1	T2	T3	T4	T5	T6	T7
Weight	0.099	0.041	0.104	0.315	0.187	0.134	0.119

Furthermore, the proof that the method was performed correctly is displayed in Table 8, in which the multiplication of the normalized matrix with the weighted matrix is calculated, allowing to subsequently calculate the consistency index (CI), the random index (RI) and consistency relationship (CR), displayed in Table 9. In this case, the consistency is less than 10%, proving that the weights were assessed correctly [34].

Table 8. Matrix multiplication.

	N × T	Priority/Weight
T1	0.785	7.89
T2	0.305	7.49
T3	0.761	7.33
T4	2.451	7.77
T5	1.516	8.10
T6	1.070	7.97
T7	0.945	7.93
Summatory		7.78

Table 9. Calculation of consistency.

Index	Value
Cl	0.13
RI	1.32
CR	0.099

3.2. VIKOR Results

In previous research, the VIKOR method was found to be the best methodology for selection of materials in automotive applications, since this delivers a compromise set of solutions as the result [35]. In this sense, Table 10 shows the results of the calculations for the VIKOR method where the normalization takes place, taking in consideration the previous weights and the best and worst parameters.

Table 10. VIKOR calculations.

Compound	Phase-Change Temp ($^\circ$C)	Density (kg/m^3)	Heat of Fusion (kJ/kg)	Specific Heat Capacity (kJ/kgK)	Thermal Conductivity (W/mK)	FOM1	FOM2
M1	0.012	0.031	0.099	0.245	0	0.034	0.119
M2	0.025	0.032	0.101	0.199	0.022	0.042	0.081
M3	0.012	0.041	0.100	0.184	0.022	0.008	0.043
M4	0.037	0.016	0.070	0.245	0.148	0	0.020
M5	0.099	0.032	0.104	0.269	0.061	0.054	0.069
M6	0.050	0.011	0.074	0.245	0.144	0.044	0.051
M7	0.025	0.041	0.100	0	0.022	0.034	0.028
M8	0.037	0.011	0.074	0.199	0.144	0.044	0.040
M9	0.037	0.016	0.102	0.170	0.079	0.094	0.092
M10	0.037	0.014	0.092	0.245	0.144	0.074	0.035
M11	0	0.012	0.093	0.210	0.151	0.083	0.026
M12	0.037	0.012	0.095	0.047	0.151	0.086	0.000
M13	0	0.011	0.001	0.184	0.151	0.115	0.024
M14	0.012	0.011	0.000	0.194	0.151	0.117	0.025
M15	0.050	0	0.064	0.315	0.187	0.134	0.039

On the other hand, the ranking is displayed in Table 11, where savENRG PCM-HS22P stands as the best, being an organic PCM that is used in the storage of great energy and which has a low cost.

Table 11. VIKOR ranking.

Compound	Si	Ri	Qi	Ranking
M1	0.541	0.245	0.607	11
M2	0.502	0.199	0.463	6
M3	0.411	0.184	0.346	3
M4	0.537	0.245	0.603	10
M5	0.688	0.269	0.797	14
M6	0.619	0.245	0.679	12
M7	0.249	0.100	0	1
M8	0.548	0.199	0.506	8
M9	0.591	0.170	0.480	7
M10	0.642	0.245	0.701	13
M11	0.575	0.210	0.557	9
M12	0.428	0.151	0.285	2
M13	0.486	0.184	0.416	4
M14	0.510	0.194	0.460	5
M15	0.790	0.315	1.000	15

Furthermore, the verification of the method says that VIKOR performed well since $Q_2 - Q_1 = 0.285$ is bigger than DQ = 0.17, showing an acceptable advantage. Furthermore, the results of S_i and R_i demonstrate that the best belongs to the winner M7, meaning that there is an acceptable stability, fulfilling the two conditions that conclude that the method has an acceptable compromised solution.

3.3. TOPSIS Results

The TOPSIS method, along with AHP, plays an important part to reduce a possible selection of a wrong PCM, which was also studied in the thermal management of electronics [36]. Moreover, the development of the normalized matrix, which takes the original criteria and divides them by the square root of the quadratic summation of all the materials, is shown in Table 12. In turn, the weighted matrix that multiplies the previous criteria by the AHP weight is displayed in Table 13, followed by the beneficial and non-beneficial solutions in Table 14. Lastly, the negative and positive ideal solutions with the closeness index for the ranking result is calculated in Table 15. In these results, again savENRG PCM-HS22P is the optimum material.

Table 12. TOPSIS normalized matrix.

Compound	Phase-Change Temp (°C)	Density (kg/m³)	Heat of Fusion (kJ/kg)	Specific Heat Capacity (kJ/kgK)	Thermal Conductivity (W/mK)	FOM1	FOM2
M1	0.275	0.335	0.214	0.231	0.440	0.312	0.401
M2	0.263	0.341	0.197	0.254	0.396	0.293	0.322
M3	0.275	0.406	0.208	0.261	0.396	0.368	0.244
M4	0.252	0.216	0.411	0.231	0.139	0.387	0.197
M5	0.195	0.340	0.180	0.220	0.316	0.267	0.298
M6	0.240	0.172	0.385	0.231	0.147	0.288	0.261
M7	0.263	0.406	0.208	0.353	0.396	0.312	0.213
M8	0.252	0.172	0.385	0.254	0.147	0.288	0.237
M9	0.252	0.212	0.191	0.268	0.279	0.177	0.346
M10	0.252	0.196	0.259	0.231	0.147	0.221	0.228
M11	0.286	0.181	0.254	0.249	0.132	0.200	0.207
M12	0.252	0.183	0.243	0.330	0.132	0.194	0.155
M13	0.286	0.175	0.169	0.261	0.132	0.129	0.204
M14	0.275	0.176	0.163	0.257	0.132	0.125	0.206
M15	0.240	0.089	0.221	0.197	0.059	0.086	0.236

Table 13. TOPSIS weighted matrix.

Compound	Phase-Change Temp (°C)	Density (kg/m³)	Heat of Fusion (kJ/kg)	Specific Heat Capacity (kJ/kgK)	Thermal Conductivity (W/mK)	FOM1	FOM2
M1	0.027	0.014	0.022	0.073	0.082	0.042	0.048
M2	0.026	0.014	0.020	0.080	0.074	0.039	0.038
M3	0.027	0.017	0.022	0.082	0.074	0.049	0.029
M4	0.025	0.009	0.043	0.073	0.026	0.052	0.023
M5	0.019	0.014	0.019	0.069	0.059	0.036	0.036
M6	0.024	0.007	0.040	0.073	0.027	0.039	0.031
M7	0.026	0.017	0.022	0.111	0.074	0.042	0.025
M8	0.025	0.007	0.040	0.080	0.027	0.039	0.028
M9	0.025	0.009	0.020	0.085	0.052	0.024	0.041
M10	0.025	0.008	0.027	0.073	0.027	0.030	0.027
M11	0.028	0.007	0.026	0.078	0.025	0.027	0.025
M12	0.025	0.007	0.025	0.104	0.025	0.026	0.018
M13	0.028	0.007	0.018	0.082	0.025	0.017	0.024
M14	0.027	0.007	0.017	0.081	0.025	0.017	0.025
M15	0.024	0.004	0.023	0.062	0.011	0.012	0.028

Table 14. Max and min TOPSIS criteria value.

Values +/−	Phase-Change Temp (°C)	Density (kg/m^3)	Heat of Fusion (kJ/kg)	Specific Heat Capacity (kJ/kgK)	Thermal Conductivity (W/mK)	FOM1	FOM2
V+	0.028	0.004	0.043	0.111	0.082	0.052	0.018
V−	0.019	0.017	0.017	0.062	0.011	0.012	0.048

Table 15. TOPSIS Ranking.

Compound	Si+	Si−	Pi	Ranking
M1	0.054	0.079	0.593	4
M2	0.047	0.073	0.608	3
M3	0.040	0.079	0.663	2
M4	0.069	0.058	0.457	8
M5	0.060	0.056	0.482	5
M6	0.070	0.045	0.394	10
M7	0.029	0.089	0.755	1
M8	0.065	0.049	0.427	9
M9	0.059	0.050	0.458	7
M10	0.073	0.037	0.335	12
M11	0.073	0.039	0.345	11
M12	0.066	0.057	0.461	6
M13	0.078	0.037	0.322	13
M14	0.079	0.036	0.312	14
M15	0.098	0.025	0.201	15

3.4. COPRAS Method

The engineering fields have been benefiting from this method by enabling the decision maker to determine the overall efficiency of the different options [37]. In this way, the process of development is displayed in Table 16 with the normalized matrix that divides each position by the summation of all of them for every criterion. Table 17 presents the weighted matrix that is the result of the previous table by every weight of AHP, followed by Table 18 that shows the summation of the beneficial and non-beneficial weights. Lastly, the priority of positions with the level of performance and the COPRAS rank is presented in Table 19, where the results show that the material M7 savENRG PCM-HS22P again is established as the better material among the 15 candidates.

Table 16. COPRAS normalized table.

Compound	Phase-Change Temp (°C)	Density (kg/m^3)	Heat of Fusion (kJ/kg)	Specific Heat Capacity (kJ/kgK)	Thermal Conductivity (W/mK)	FOM1	FOM2
M1	0.071	0.093	0.058	0.060	0.130	0.085	0.107
M2	0.068	0.095	0.053	0.066	0.117	0.080	0.086
M3	0.071	0.113	0.056	0.068	0.117	0.101	0.065
M4	0.065	0.060	0.111	0.060	0.041	0.106	0.052
M5	0.050	0.094	0.049	0.057	0.093	0.073	0.079
M6	0.062	0.048	0.104	0.060	0.043	0.079	0.069
M7	0.068	0.113	0.056	0.092	0.117	0.085	0.057
M8	0.065	0.048	0.104	0.066	0.043	0.079	0.063
M9	0.065	0.059	0.052	0.070	0.082	0.048	0.092
M10	0.065	0.055	0.070	0.060	0.043	0.061	0.061
M11	0.074	0.050	0.069	0.065	0.039	0.055	0.055
M12	0.065	0.051	0.066	0.086	0.039	0.053	0.041
M13	0.074	0.049	0.046	0.068	0.039	0.035	0.054
M14	0.071	0.049	0.044	0.067	0.039	0.034	0.055
M15	0.062	0.025	0.060	0.051	0.017	0.024	0.063

Table 17. COPRAS weighted matrix.

Compound	Phase-Change Temp (°C)	Density (kg/m^3)	Heat of Fusion (kJ/kg)	Specific Heat Capacity (kJ/kgK)	Thermal Conductivity (W/mK)	FOM1	FOM2
M1	0.0071	0.0038	0.0060	0.0190	0.0243	0.0115	0.0127
M2	0.0068	0.0039	0.0055	0.0209	0.0219	0.0108	0.0102
M3	0.0071	0.0046	0.0059	0.0215	0.0219	0.0136	0.0078
M4	0.0065	0.0024	0.0116	0.0190	0.0077	0.0143	0.0062
M5	0.0050	0.0038	0.0051	0.0181	0.0174	0.0098	0.0095
M6	0.0062	0.0019	0.0108	0.0190	0.0081	0.0106	0.0083
M7	0.0068	0.0046	0.0059	0.0290	0.0219	0.0115	0.0068
M8	0.0065	0.0019	0.0108	0.0209	0.0081	0.0106	0.0075
M9	0.0065	0.0024	0.0054	0.0221	0.0154	0.0065	0.0110
M10	0.0065	0.0022	0.0073	0.0190	0.0081	0.0082	0.0072
M11	0.0074	0.0020	0.0072	0.0205	0.0073	0.0074	0.0066
M12	0.0065	0.0021	0.0068	0.0271	0.0073	0.0071	0.0049
M13	0.0074	0.0020	0.0048	0.0215	0.0073	0.0047	0.0065
M14	0.0071	0.0020	0.0046	0.0211	0.0073	0.0046	0.0065
M15	0.0062	0.0010	0.0062	0.0162	0.0032	0.0032	0.0075

Table 18. COPRAS solutions.

Compound	S^{+i}	S^{-i}
M1	0.068	0.017
M2	0.066	0.014
M3	0.070	0.012
M4	0.059	0.009
M5	0.055	0.013
M6	0.055	0.010
M7	0.075	0.011
M8	0.057	0.009
M9	0.056	0.013
M10	0.049	0.009
M11	0.050	0.009
M12	0.055	0.007
M13	0.046	0.008
M14	0.045	0.009
M15	0.035	0.009

Table 19. COPRAS ranking.

Compound	Qi	Ui	Rank
M1	0.074	88%	3
M2	0.074	87%	4
M3	0.079	93%	2
M4	0.071	85%	5
M5	0.064	75%	10
M6	0.065	77%	8
M7	0.085	100%	1
M8	0.068	81%	7
M9	0.064	76%	9
M10	0.060	72%	12
M11	0.062	74%	11
M12	0.070	83%	6
M13	0.058	69%	13
M14	0.057	68%	14
M15	0.048	56%	15

3.5. Spearman's Correlation Results

Table 20 shows the results of the Spearman's correlations where only the correlation COPRAS-TOPSIS has a very good relation; thus, the rest of them does not have a good relation although they indicate a positive correlation. However, it can be seen that the three methods show consistency in their results regarding the best material, as studied before [27].

Table 20. Spearman's correlation.

	Correlation of 15 Materials Ranked		
	COPRAS	TOPSIS	VIKOR
COPRAS	-	0.914	0.439
TOPSIS	-	-	0.407

Moreover, even though the correlation shows that it is not perfect, it agrees with the fact that the material savENRG PCM-HS22P was chosen as the best, and since it is the one that will be simulated, Table 21 displays its properties.

Table 21. Selected material for simulation.

Compound	Phase-Change Temp (°C)	Density (kg/m^3)	Heat of Fusion (kJ/kg)	Specific Heat Capacity (kJ/kgK)	Thermal Conductivity (W/mK)
savENRG PCM-HS22P	23	1540	185	3.05	0.54

Comparing the selection results with what Rastogi et al. had done, there is a difference where they prioritized the Figures of Merit since air conditioning systems require that the extraction of heat be primordial [25]. On the other hand, this application demands heat storage and release. Moreover, in the research of Socaciu et al., the best materials come from the same family as our best (SavEnrg PCM-Hs22P), also in thermal comfort applications in the automotive industry [38], corroborating that our selection is optimum.

3.6. Simulation Results

By reducing the difference in temperature between the internal air and body heat, the thermal comfort of the user can be controlled [39]. In this sense, the simulation with a raising temperature in the roof showed that without PCMs there is not much difference in the internal air, resulting in a high temperature in the cabin, as shown in Figure 2a. However, by using a 20 mm layer of the selected PCM, the temperature did not rise as much, and presented a decrease of near 9 °C, which is a congruent result compared with previous research that managed almost the same temperature degree with other PCMs [40–42]. This difference is showed in Figure 2b.

The discharging event, on the other hand, showed that while the temperature of the rooftop dropped to 12 °C, the internal air stood 4 °C warmer, meaning that the PCM maintain the internal comfort by releasing its stored energy into the cabin; this result is displayed in Figure 3.

Other researchers proved that the utilization of PCMs delayed the heat flux, making them more effective for isolation since the wall loses less energy, which helps to maintain a better temperature [9,43].

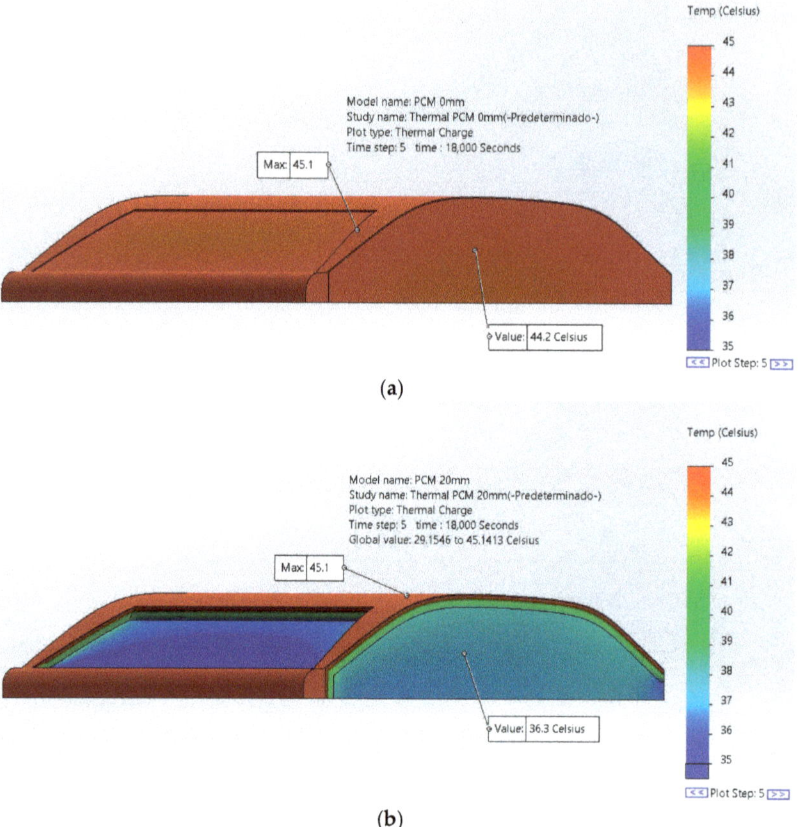

(a)

(b)

Figure 2. Charging the simulation comparison. (**a**) PCM 0 mm (**b**) PCM 20 mm.

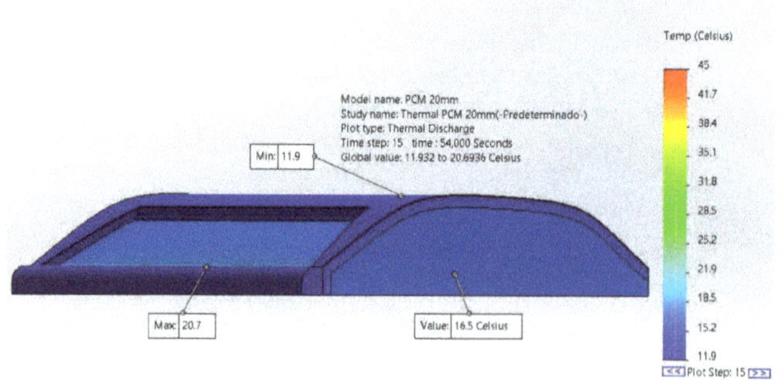

Figure 3. Discharge.

4. Conclusions

Starting with a list of 15 PCMs, savENRG PCM-HS22P was selected as the best by the utilization of the AHP, VIKOR, COPRAS and TOPSIS multicriteria methods. Furthermore, a Spearman's correlation showed that the methods are consistent, and the selection is optimum.

The simulation showed that the material savENRG PCM-HS22P improves the thermal comfort in the cabin by reducing the internal temperature of the air by 9 °C in the charging event, and in the discharging simulation it was also showed that the PCM allows to maintain a better temperature inside the cabin by making a differentiation of 4 °C warmer, both with a 20 mm layer of the selected PCM.

It is demonstrated that the use of a PCM layer of 20 mm can improve the thermal comfort in the vehicle, reducing the need to use the heater or the air conditioner.

Author Contributions: Conceptualization: J.M.-G. and J.F.N.; methodology: M.C., J.M.-G. and J.F.N.; software: M.C., J.M.-G. and J.F.N.; validation: M.C., J.M.-G., J.F.N., R.B.S. and E.R.; formal analysis: E.R., J.F.N. and J.M.-G.; investigation: E.R., J.F.N. and J.M.-G.; resources: E.R.; writing—original draft preparation: J.F.N.; writing—review and editing: E.R., R.B.S. and J.M.-G.; visualization: J.F.N.; supervision: J.M.-G.; project administration: J.M.-G.; funding acquisition: J.M.-G. All authors have read and agreed to the published version of the manuscript.

Funding: This research takes part of the project 'Selection, characterization and simulation of phase change materials for thermal comfort, cooling and energy storage'. This project is part of the INEDITA call for R&D research projects in the field of energy and materials. This research is part of the project P121819, Parque de Energias Renovables, founded by Universidad International SEK.

Institutional Review Board Statement: Not applicable.

Informed Consent Statement: Not applicable.

Data Availability Statement: The data reported in this research can be found at https://drive.google.com/file/d/19rI2-2r6nYN0C3Ymjiegf55bdq4jrB8Z/view?usp=sharing (accessed on 15 January 2021).

Conflicts of Interest: The authors declare that the research was conducted in the absence of any commercial or financial relationships that could be construed as a potential conflict of interest.

References

1. Farrington, R.; Rugh, J. Impact of Vehicle Air-Conditioning on Fuel Economy, Tailpipe Emissions, and Electric Vehicle Range: Preprint. 2000. Available online: https://www.osti.gov/biblio/764573 (accessed on 15 January 2021).
2. Antonijevic, D.; Heckt, R. Heat pump supplemental heating system for motor vehicles. *Proc. Inst. Mech. Eng. Part D J. Automob. Eng.* **2004**, *218*, 1111–1115. [CrossRef]
3. Kang, B.H.; Lee, H.J. A Review of Recent Research on Automotive HVAC Systems for EVs. *Int. J. Air-Cond. Refrig.* **2017**, *25*, 1730003. [CrossRef]
4. Simion, M.; Socaciu, L.; Unguresan, P. Factors which Influence the Thermal Comfort Inside of Vehicles. *Energy Procedia* **2016**, *85*, 472–480. [CrossRef]
5. Prajapati, D.G.; Kandasubramanian, B. Biodegradable Polymeric Solid Framework-Based Organic Phase-Change Materials for Thermal Energy Storage. *Ind. Eng. Chem. Res.* **2019**, *58*, 10652–10677. [CrossRef]
6. Jaguemont, J.; Omar, N.; Van den Bossche, P.; Mierlo, J. Phase-change materials (PCM) for automotive applications: A review. *Appl. Therm. Eng.* **2018**, *132*, 308–320. [CrossRef]
7. Nazir, H.; Batool, M.; Bolivar Osorio, F.J.; Isaza-Ruiz, M.; Xu, X.; Vignarooban, K.; Phelan, P.; Inamuddin; Kannan, A.M. Recent developments in phase change materials for energy storage applications: A review. *Int. J. Heat Mass Transf.* **2019**, *129*, 491–523. [CrossRef]
8. Kuznik, F.; Virgone, J. Experimental assessment of a phase change material for wall building use. *Appl. Energy* **2009**, *86*, 2038–2046. [CrossRef]
9. Liu, H.; Awbi, H.B. Performance of phase change material boards under natural convection. *Build. Environ.* **2009**, *44*, 1788–1793. [CrossRef]
10. Zhang, N.; Yuan, Y.; Cao, X.; Du, Y.; Zhang, Z.; Gui, Y. Latent Heat Thermal Energy Storage Systems with Solid–Liquid Phase Change Materials: A Review. *Adv. Eng. Mater.* **2018**, *20*, 1–30. [CrossRef]
11. Bakan, G.; Gerislioglu, B.; Dirisaglik, F.; Jurado, Z.; Sullivan, L.; Dana, A.; Lam, C.; Gokirmak, A.; Silva, H. Extracting the temperature distribution on a phase-change memory cell during crystallization. *J. Appl. Phys.* **2016**, *120*, 164504. [CrossRef]
12. Gerislioglu, B.; Bakan, G.; Ahuja, R.; Adam, J.; Mishra, Y.K.; Ahmadivand, A. The role of $Ge_2Sb_2Te_5$ in enhancing the performance of functional plasmonic devices. *Mater. Today Phys.* **2020**, *12*, 100178. [CrossRef]
13. Kuznik, F.; David, D.; Johannes, K.; Roux, J.J. A review on phase change materials integrated in building walls. *Renew. Sustain. Energy Rev.* **2011**, *15*, 379–391. [CrossRef]
14. Zhou, D.; Zhao, C.Y.; Tian, Y. Review on thermal energy storage with phase change materials (PCMs) in building applications. *Appl. Energy* **2012**, *92*, 593–605. [CrossRef]

15. Kim, K.; Choi, K.; Kim, Y.; Lee, K.; Lee, K. Feasibility study on a novel cooling technique using a phase change material in an automotive engine. *Energy* **2010**, *35*, 478–484. [CrossRef]
16. Yu, X.; Li, Z.; Lu, Y.; Huang, R.; Roskilly, A.P. Investigation of organic Rankine cycle integrated with double latent thermal energy storage for engine waste heat recovery. *Energy* **2019**, *170*, 1098–1112. [CrossRef]
17. Yang, K.; Zhu, N.; Chang, C.; Wang, D.; Yang, S.; Ma, S. A methodological concept for phase change material selection based on multi-criteria decision making (MCDM): A case study. *Energy* **2018**, *165*, 1085–1096. [CrossRef]
18. Amer, A.E.; Rahmani, K.; Lebedev, V.A. Using the Analytic Hierarchy Process ({AHP}) method for selection of phase change materials for solar energy storage applications. *J. Phys. Conf. Ser.* **2020**, *1614*, 12022. [CrossRef]
19. Mukhamet, T.; Kobeyev, S.; Nadeem, A.; Memon, S.A. Ranking PCMs for building façade applications using multi-criteria decision-making tools combined with energy simulations. *Energy* **2021**, *215*, 119102. [CrossRef]
20. Méndez, A.; Martínez-Gómez, J.; Rodríguez, F.; Nicolalde, J.F. Selección de un material de cambio de fase mediante el uso del método de selección multicriterio para su uso en un sistema de almacenamiento térmico automotriz. *RISTI Rev. Iber. Sist. Tecnol. Inf.* **2020**, 113–125. Available online: https://www.proquest.com/openview/63deba1836c13b0f5ceb448dbcbfa2e3/1?pq-origsite=gscholar&cbl=1006393 (accessed on 15 January 2021).
21. Amoozad Mahdiraji, H.; Arzaghi, S.; Stauskis, G.; Zavadskas, E.K. A Hybrid Fuzzy BWM-COPRAS Method for Analyzing Key Factors of Sustainable Architecture. *Sustainability* **2018**, *10*, 1626. [CrossRef]
22. Shojaeefard, M.H.; Khalkhali, A.; Miandoabchi, E. Multi-criteria decision making approach for selecting the best friction distribution in superplastic forming of a vehicle component. *Proc. Inst. Mech. Eng. Part E J. Process Mech. Eng.* **2014**, *230*, 146–157. [CrossRef]
23. Singh, R.P.; Xu, H.; Kaushik, S.C.; Rakshit, D.; Romagnoli, A. Charging performance evaluation of finned conical thermal storage system encapsulated with nano-enhanced phase change material. *Appl. Therm. Eng.* **2019**, *151*, 176–190. [CrossRef]
24. Dhir, D.K. Thermo-mechanical performance of automotive disc brakes. *Mater. Today Proc.* **2018**, *5*, 1864–1871. [CrossRef]
25. Rastogi, M.; Chauhan, A.; Vaish, R.; Kishan, A. Selection and performance assessment of Phase Change Materials for heating, ventilation and air-conditioning applications. *Energy Convers. Manag.* **2015**, *89*, 260–269. [CrossRef]
26. Odu, G.O. Weighting methods for multi-criteria decision making technique. *J. Appl. Sci. Environ. Manag.* **2019**, *23*, 1449. [CrossRef]
27. Shekhovtsov, A.; Salabun, W. A comparative case study of the VIKOR and TOPSIS rankings similarity. *Procedia Comput. Sci.* **2020**, *176*, 3730–3740. [CrossRef]
28. Beltrán, R.D.; Martínez-Gómez, J. Analysis of phase change materials (PCM) for building wallboards based on the effect of environment. *J. Build. Eng.* **2019**, *24*, 100726. [CrossRef]
29. Mousavi-nasab, S.H.; Sotoudeh-anvai, A. A comprehensive MCDM-based approach using TOPSIS, COPRAS and DEA as an auxiliary tool for material selection problems. *Mater. Des.* **2017**, *121*, 237–253. [CrossRef]
30. Mousavi-Nasab, S.H.; Sotoudeh-Anvari, A. A new multi-criteria decision making approach for sustainable material selection problem: A critical study on rank reversal problem. *J. Clean. Prod.* **2018**, *182*, 466–484. [CrossRef]
31. Granta Design Limited. *CES-Edupack*; Granta Design Limited: Cambridge, UK, 2019.
32. Dadour, I.R.; Almanjahie, I.; Fowkes, N.D.; Keady, G.; Vijayan, K. Temperature variations in a parked vehicle. *Forensic Sci. Int.* **2011**, *207*, 205–211. [CrossRef]
33. Cengel, Y.; Boles, M.A. *Termodinámica*, 8th ed.; McGraw-Hill Interamericana: New York, NY, USA, 2015.
34. Nadeem, A.; Rakhman, K.; Hossain, M.A. Phase Change Materials Ranking by Using the Analytic Hierarchy Process. *Proc. Int. Struct. Eng. Constr.* **2020**, *7*, CPM-09-1–CPM-09-6. [CrossRef]
35. Jeya Girubha, R.; Vinodh, S. Application of fuzzy VIKOR and environmental impact analysis for material selection of an automotive component. *Mater. Des.* **2012**, *37*, 478–486. [CrossRef]
36. Kumar, A.; Kothari, R.; Sahu, S.K.; Kundalwal, S.I. Selection of phase-change material for thermal management of electronic devices using multi-attribute decision-making technique. *Int. J. Energy Res.* **2021**, *45*, 2023–2042. [CrossRef]
37. Chatterjee, P.; Athawale, V.M.; Chakraborty, S. Materials selection using complex proportional assessment and evaluation of mixed data methods. *Mater. Des.* **2011**, *32*, 851–860. [CrossRef]
38. Socaciu, L.; Giurgiu, O.; Banyai, D.; Simion, M. PCM selection using AHP method to maintain thermal comfort of the vehicle occupants. *Energy Procedia* **2016**, *85*, 489–497. [CrossRef]
39. Holmer, I.; Nilsson, H.; Bohm, M.; Noren, O. Thermal Aspects of Vehicle Comfort. *Appl. Hum. Sci.* **1995**, *14*, 159–165. [CrossRef]
40. Purusothaman, M.; Saichand, K.; Sam Cornilius, C.; Siva, R. Experimental Investigation of Thermal Performance in a Vehicle Cabin Test Setup with Pcm in the Roof. In *IOP Conference Series: Materials Science and Engineering*; IOP Publishing: Bristol, UK, 2017. [CrossRef]
41. Saleel, C.A.; Mujeebu, M.A.; Algarni, S. Coconut oil as phase change material to maintain thermal comfort in passenger vehicles. *J. Therm. Anal. Calorim.* **2019**, *136*, 629–636. [CrossRef]
42. Oró, E.; de Jong, E.; Cabeza, L.F. Experimental analysis of a car incorporating phase change material. *J. Energy Storage* **2016**, *7*, 131–135. [CrossRef]
43. Arıcı, M.; Bilgin, F.; Nižetić, S.; Karabay, H. PCM integrated to external building walls: An optimization study on maximum activation of latent heat. *Appl. Therm. Eng.* **2020**, *165*, 114560. [CrossRef]

 *applied sciences*

Review

Towards Phase Change Materials for Thermal Energy Storage: Classification, Improvements and Applications in the Building Sector

Christina V. Podara, Ioannis A. Kartsonakis *[iD] and Costas A. Charitidis *[iD]

Research Unit of Advanced, Composite, Nano-Materials and Nanotechnology, School of Chemical Engineering, National Technical University of Athens, GR-15773 Athens, Greece; christina.podara@hotmail.com
* Correspondence: ikartso@chemeng.ntua.gr (I.A.K.); charitidis@chemeng.ntua.gr (C.A.C.);
  Tel.: +30-210-772-4046 (C.A.C.)

Abstract: The management of energy consumption in the building sector is of crucial concern for modern societies. Fossil fuels' reduced availability, along with the environmental implications they cause, emphasize the necessity for the development of new technologies using renewable energy resources. Taking into account the growing resource shortages, as well as the ongoing deterioration of the environment, the building energy performance improvement using phase change materials (PCMs) is considered as a solution that could balance the energy supply together with the corresponding demand. Thermal energy storage systems with PCMs have been investigated for several building applications as they constitute a promising and sustainable method for reduction of fuel and electrical energy consumption, while maintaining a comfortable environment in the building envelope. These compounds can be incorporated into building construction materials and provide passive thermal sufficiency, or they can be used in heating, ventilation, and air conditioning systems, domestic hot water applications, etc. This study presents the principles of latent heat thermal energy storage systems with PCMs. Furthermore, the materials that can be used as PCMs, together with the most effective methods for improving their thermal performance, as well as various passive applications in the building sector, are also highlighted. Finally, special attention is given to the encapsulated PCMs that are composed of the core material, which is the PCM, and the shell material, which can be inorganic or organic, and their utilization inside constructional materials.

Keywords: phase change materials; thermal energy storage; energy efficiency; building applications; construction materials

Citation: Podara, C.V.; Kartsonakis, I.A.; Charitidis, C.A. Towards Phase Change Materials for Thermal Energy Storage: Classification, Improvements and Applications in the Building Sector. *Appl. Sci.* **2021**, *11*, 1490. https://doi.org/10.3390/app11041490

Academic Editor: Cesare Biserni
Received: 23 December 2020
Accepted: 3 February 2021
Published: 6 February 2021

Publisher's Note: MDPI stays neutral with regard to jurisdictional claims in published maps and institutional affiliations.

1. Introduction

The contemporary societies have enhanced energy needs, leading to an increasingly intensive research for the development of energy storage technologies. Global energy consumption, along with CO_2 and greenhouse gasses emissions, is accelerating at a very fast pace due to global population growth, rapid global economic growth, and the ever-increasing human dependence on energy-consuming appliances. This rapid increase in global energy consumption has particular environmental implications that pose serious challenges to human health and the environment. Another consequence of the increment of global energy consumption is the reduction in the availability of traditional energy resources, such as oil, coal, and natural gas. This outcome has created a growing need for the development of new systems for the conversion and storage of clean and sustainable energy [1,2].

The building sector has a very large impact on the overall global energy consumption, and, furthermore, to the environment, by high emissions of greenhouse gasses, CO_2, and the acidification of the oceans. The increased energy consumption does not occur only during the production of structural materials and the building construction, but mostly for the operation of the premises [3]. It has been proven that buildings' operation energy consumption for providing indoor comfort accounts for 30% of total energy consumption [4]. Thermal energy storage (TES) is a promising and sustainable method for decreasing the energy consumptions in the building sector. Systems of TES using phase change materials (PCMs) find numerous applications for providing and maintaining a comfortable environment of the building envelope, without consumption of electrical energy or fuel [5].

Phase change materials are substances that are able to absorb and store large amounts of thermal energy. The mechanism of PCMs for energy storage relies on the increased energy need of some materials to undergo phase transition. They are able to absorb sensible heat as their temperature rise, and, at the phase change temperature, absorb a large amount of heat, which is called latent heat of fusion, in order to change phase. The energy stays stored in the PCM until the temperature decreases and the material undergoes phase transition again, which also signifies the energy release [1].

The materials used as PCMs can be classified based on the type of phase change to solid-liquid, liquid-gas, and solid-solid compounds. The latent heat in solid-solid PCMs, such as polyurethanes, cross-linked polyethylene, and other polymers, is relatively low compared to the latent heat of those that are in the solid-liquid form [6]. As an outcome, the PCMs used for building applications are mostly solid-liquid type. These can be organic, such as paraffin, fatty acids, etc., inorganic, such as hydrate salts and metals, or their eutectic mixtures. Organic PCMs are the best candidates for multiple building applications, though their low thermal conductivity requires the fabrication of composite materials with enhanced thermal conductivity. The most investigated approach is the encapsulation of PCMs with thermal conductive materials. The encapsulation method presents many advantages, such as the leakage of liquid phase management and the protection of the PCM from corrosion. Another approach is the formation of form-stable composite PCMs, which consist of the PCM dispersed inside a cross-linked polymer matrix, a porous mineral material or expanded graphite or perlite [3,5].

There has been a lot of research for the utilization of these materials in active and passive building applications for the reduction of electrical energy and fuel consumption. Phase change materials are used in active applications for heating, ventilation, and air conditioning systems, domestic hot water applications, suspended ceilings, external solar facades, etc. [4]. Furthermore, PCMs are used in passive applications in buildings that aim for the improvement of the thermal performance of the construction systems such as the ceiling, the walls, and the floor. These applications include the incorporation of the PCMs inside constructional materials [7].

2. Thermal Energy Storage (TES)

Energy storage systems aim for the conversion of energy into a form that can be stored in order to be used when there is necessity. Thermal energy storage system is a type of a sustainable energy storage system that is based on the utilization of materials that can store thermal energy when increasing their temperature and release it when the temperature is reduced. There are three types of TES systems: sensible heat, latent heat, and chemical storage system (Figure 1). The present work presents latent heat storage systems using PCMs [8,9].

2.1. Sensible Heat Storage (SHS)

In TES systems, thermal energy can be stored either as sensible heat or as latent heat (Figure 2). In case of sensible heat storage (SHS) systems, storing of energy is induced by utilization of the heat capacity gained by temperature increment of the material. The energy storage capacity of SHS systems depends on the specific heat capacity of the material, the quantity of the material and the temperature change gradient. Radiation, convection, or conduction are the parameters that affect and can raise the corresponding temperature [9,10].

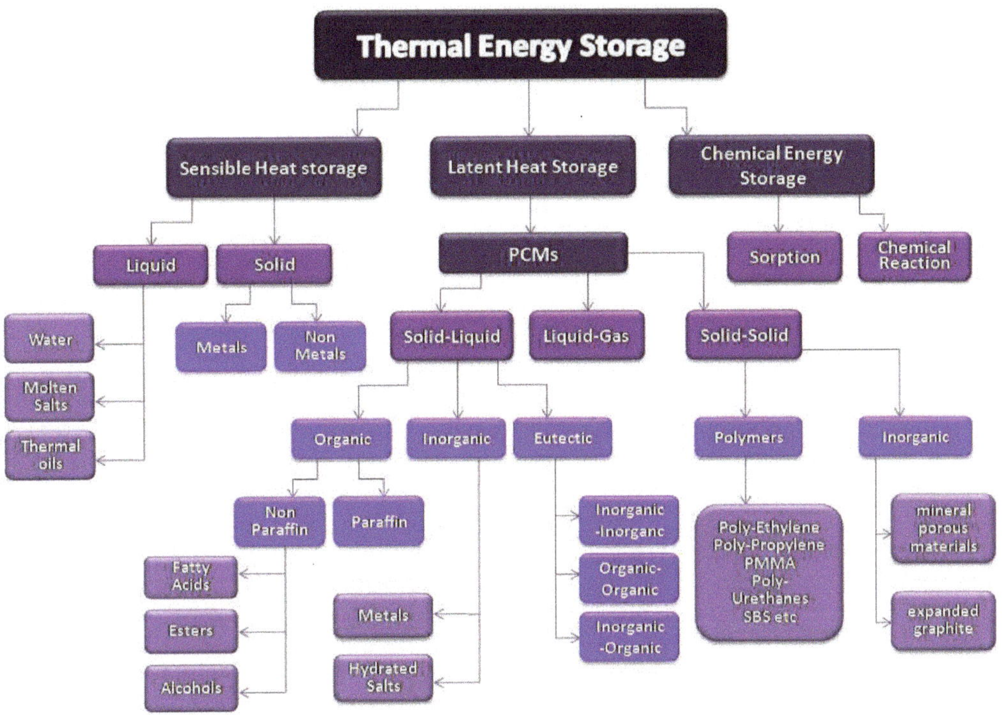

Figure 1. Classification of thermal energy storage types and materials.

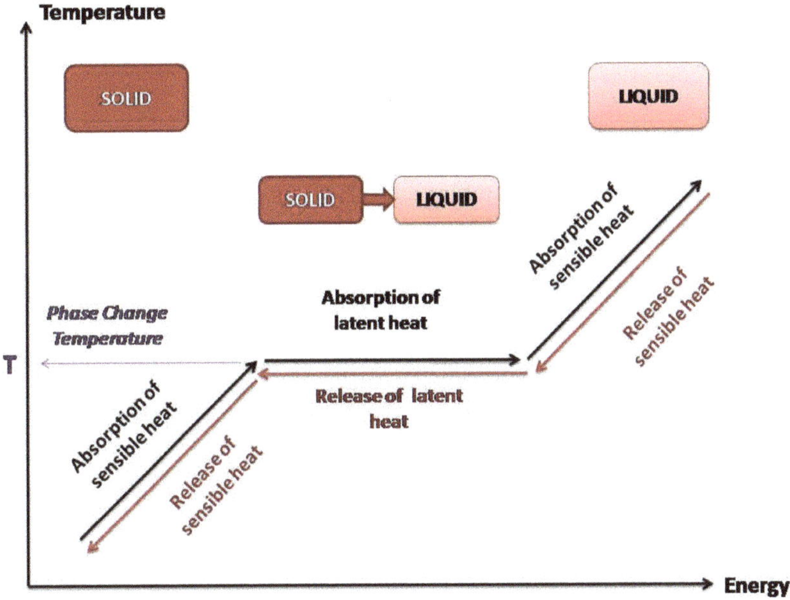

Figure 2. Sensible and latent heat.

The materials used for SHS are either in the liquid phase or the solid phase. The utilized liquid phase materials are water, molten salts, and oils. Water as an SHS material is very efficient for applications in temperatures below 100 °C, due to its high specific heat capacity, abundance, and low cost. Molten salts also have desired properties, and they are suitable for application above 100 °C. On the other hand, oils have low specific heat capacity and thermal conductivity, thus their performance is limited. Solid-phase materials used for SHS can be metals or nonmetals. Metals such as aluminum, copper, iron, and metal alloys have high thermal conductivity and specific heat capacity, though their high cost limits their application. Nonmetallic materials that can be used for SHS are rocks, concrete, marbles, bricks, granite, etc., but the disadvantage of all of these materials lies in their low thermal conductivity and specific heat capacity [6,8].

2.2. Latent Heat Storage

Latent heat storage (LHS) systems are based on the absorption or release of heat that takes place during the phase transition of a material (gas to liquid, liquid to solid, and vice versa). The amount of energy being stored depends on the mass of the material used and the latent heat of fusion, which is characteristic for every material, and it is related to its molecular structure. The materials used for LHS are also PCMs due to the fact that they store energy through phase transition. The energy storage consists of both sensible and latent heat, as presented in Figure 2. The increase of temperature to the phase change temperature results in the absorption of sensible heat from the PCM. At phase change temperature, the PCM absorbs latent heat at a molecular level for the phase transition. This amount of heat is called latent heat of fusion or evaporation, depending on the kind of phase change. The systems of LHS that use PCMs are considered as a promising thermal storage technology and they have been investigated in a large range for many applications [6,8,9].

2.3. Thermal Energy Storage in Buildings

There is urgent necessity for the development of new technologies for energy conservation in the building sector. The application of TES systems is a promising technology with many advantages, including the increasing of energy and economic efficiency in buildings, as well as the reduction of electrical energy consumption and, furthermore, the reduction of environmental pollution and CO_2 emissions [11].

The systems of TES that are applied in buildings can be either active or passive. Active TES systems are characterized by forced convection heat transfer and, in some cases, also mass transfer, such as a heat exchanger. Active systems mainly aim to control the indoor conditions of a building, for example providing free cooling. Additionally, an active TES system can be applied for peak shaving purposes in heating, ventilation, and air conditioning systems [7].

On the other hand, passive TES systems aim to create or maintain comfort conditions in the building, exploiting naturally available energy sources such as the solar energy or the wind, along with architectural design of building components, such as shading effect using blinds, etc. Passive TES systems minimize the use of mechanically-assisted heating or cooling systems. The main idea is to maximize the comfort conditions of the building envelope without using external energy sources. Some examples of passive TES systems are the use of ventilation facades and the PCM wallboards [12,13].

3. Phase Change Materials

3.1. Classification of PCMs

There is a wide variety of compounds that are used as phase change materials for numerous different applications due to their different physical and chemical properties. For building applications, solid-liquid PCMs are preferred because of their ability to absorb and release large amounts of energy within a narrow range of temperature. Solid-liquid

PCMs used in buildings can be classified into three major groups: organic compounds, inorganic compounds, and eutectics of organic and/or inorganic compounds.

3.1.1. Organic Materials

Organic PCMs can be paraffinic, such as paraffin wax, which is the most commercially used organic PCM for building applications, or they can be nonparaffinic, such as fatty acids, fatty acid esters, glycolic acids, alcohols, etc. Paraffinic PCMs consist of hydrocarbon chains, while nonparaffinic also contain electronegative atoms in their chains [8,9]. Paraffin waxes consist of hydrocarbon chains ranging from 8 to 15 carbon atoms, and pure paraffin of hydrocarbon chains range from 14 to 40 carbon atoms. Thermal properties of some paraffin waxes and pure paraffins are presented in Table 1.

Table 1. Thermal properties of some paraffins.

Material	Chemicalformula	Melting Temperature (°C)	Melting Enthalpy (J/g)	Thermal Conductivity (W/(m.K))	Density (kg/m^3)	Ref
n-Tetradecane	$C_{14}H_{30}$	5.5	228	-	-	[8]
n-Pentadecane	$C_{15}H_{32}$	10	205	-	-	[8]
n-Hexadecane	$C_{16}H_{34}$	18.0	210.0–238.0	0.2 (solid)	760.0 (liquid, 20.0 °C)	[10]
n-Heptadecane	$C_{17}H_{36}$	19.0	240.0	0.2	776.0 (liquid, 20.0 °C)	[10]
n-Octadecane	$C_{18}H_{38}$	28.0	200.0–245.0	0.15 (liquid, 40.0 °C) 0.36 (solid, 25.0 °C)	774.0 (liquid, 70.0 °C) 814.0 (solid, 20.0 °C)	[10]
		28.0	179	0.2	750.0 (liquid) 870.0 (solid)	[4]
n-Nonadecane	$C_{19}H_{40}$	32.0–33.0	222.0	0.18 (liquid, 60.0 °C) 0.26 (solid, 19.0 °C)	780.0	[10]
n- Eicosane	$C_{20}H_{42}$	36.0–37.0	247	0.15 (liquid) 0.42 (solid)	-	[10]
n-Heneicosane	$C_{21}H_{44}$	39.0–41.0	201.0	-	-	[10]
n-Docosane	$C_{22}H_{46}$	44.0	249	-	-	[8]
n-Tricosane	$C_{23}H_{48}$	47.5	232	-	-	[8]
n-Tetracosane	$C_{24}H_{50}$	50.6	255	-	-	[8]

Nonparaffinic PCMs can be fabricated from biosourced raw materials, such as animal fats, and vegetable oils, such as beef tallow, margarine, coconut oil, castor oil, etc. These materials are sustainable, biodegradable, nontoxic, and less flammable. On the other hand, their major drawbacks are that they are more corrosive than paraffin and more expensive [14,15]. Some nonparaffinic PCMs are presented in Table 2.

Table 2. Thermal properties of nonparaffins.

	Material	Melting Temperature (°C)	Melting Enthalpy (J/g)	Ref
Fatty acids	Capric Acid	30.2	142.7	[4]
	Capric Acid	32	152.7	[5]
	Lauric acid	43.05	172.3	[5]
	Myristic acid	51.80	178.14	[13]
	Palmitic acid	60.42	233.24	[13]
	Stearic acid	54.29	188.28	[5]
Fatty acids esters	Butyl stearate	19	140	[16]
	Propyl palmitate	19	186	[17]
	Methyl palmitate	29	205	[8]
	Methyl eicosanate	45	230	[8]
	Methyl behenate	52	234	[8]
Alcohols	1-dodecanol	26	200	[17]
	Phenol	41	120	[8]
	Cetyl alcohol	49.3	141	[8]
Polyethylene glycol (PEG)	PEG 400	3.2	91.4	[13]
	PEG 600	22.2	108.4	[13]
	PEG800	25.39	133.6	[5]
	PEG 1000	34.89	143.62	[13]
	PEG 2000	52.63	180.70	[13]
	PEG 4000	48.95	183.10	[13]

Organic PCMs exhibit a great number of advantages. First of all, the phase change temperature rises in proportion to the number of carbon atoms in the chain, providing availability in a broad range of temperatures for different applications. Another advantage of this type of PCMs is that they are noncorrosive, chemically inert, and thermally stable in the temperature range needed, therefore, no degradation problems occur during their use. Additionally, they demonstrate a congruent melting process which means that the phase change happens repeatedly without phase segregation. Furthermore, they have high latent heat, good nucleation properties, low liquid phase sub-cooling capability, and minimal volume variation. Finally, they are compatible with most of the construction materials, they can be recycled, and their cost is relatively low. On the other hand, organic PCMs also exhibit some disadvantages based on their low energy storage capacity, low density, low thermal conductivity, and high flammability [8,9,14,15].

3.1.2. Inorganic Materials

There are two kinds of inorganic compounds that can be used as PCMs: hydrated salts and metals. Hydrated salts have been immensely studied for their use as PCMs. Their general formula is $AB \cdot nH_2O$, and the phase change is in fact the dehydration reaction of the hydrated salt. Regarding metal category, they include low melting metals (eutectics) but these materials have not yet been seriously studied as PCMs [8]. The greatest properties of inorganic PCMs are the high thermal conductivity and the high energy efficiency (high enthalpy). In addition, other advantages are the low volume change during phase transition and that they are inflammable [14].

The main drawback of inorganic PCMs is the incongruent melting which leads to phase segregation and, furthermore, to decreased performance after every charge-discharge circle. Another disadvantage is the poor nucleating properties that resulting in supercooling of the liquid phase before crystallization. Additionally, inorganic PCMs are widely corrosive, in some cases toxic, show limited compatibility with construction materials, and they are rather expensive [14,15]. The advantages and disadvantages for both organic and inorganic PCMs are summarized in Table 3.

Table 3. Advantages and disadvantages of organic and inorganic phase change materials (PCMs).

	Advantages	Disadvantages
Organic PCMs	Wide phase change temperature range. Thermally stable—no degradation. Chemically inert. Noncorrosive. Congruent melting process—no phase segregation. High latent heat. Good nucleation properties. Low liquid phase sub-cooling capability. Minimal volume variation. Compatible with most of the construction materials. Recyclable. Low cost.	Low thermal conductivity. Low density. High flammability.
Inorganic PCMs	High thermal conductivity. High energy efficiency (high enthalpy). Low volume change during phase transition. Non-flammable.	Incongruent melting—phase segregation. Poor nucleating. Supercooling of the liquid phase. Corrosiveness. Toxicity. Limited compatibility with construction materials. Higher cost.

3.1.3. Eutectic Mixtures

The eutectic accrues from the mixture of two or more compounds of organic, inorganic, or both PCMs. In this way, an immeasurable number of eutectic mixtures can be fabricated. This gives the opportunity to produce a material with the desiring thermal properties, such as phase change temperature and latent heat. So, the most important advantage of eutectic mixtures is the ability of producing a material with the suitable phase change temperature for an application. The organic eutectic mixtures show some extra advantages in comparison with inorganic or inorganic/organic mixtures. These are the long-term stability, high phase change enthalpy, and facility of impregnating into porous support material due to their high chemical compatibility and surface tension. The main disadvantage is their high cost [14,18]. Thermal properties of some hydrated salts and eutectic mixtures are presented in Table 4.

Table 4. Thermal properties of inorganic materials and eutectic mixtures used as PCMs.

Inorganic Material	Melting Temperature (°C)	Melting Enthalpy (J/g)	Ref	Eutectic Mixtures	Melting Temperature (°C)	Melting Enthalpy (J/g)	Ref
$CaCl_2 \cdot 12H_2O$	29.8	174	[8]	Capric Acid-Palmitic Acid	26.2	177	[4]
$LiNO_3 \cdot H_2O$	30.0	296	[8]	Capric Acid-Myristic Acid	21.7	155	[4]
$LiNO_3 \cdot 3H_2O$	30	189	[8]	Capric Acid-Stearic Acid	24.7	179	[4]
$LiNO_3 \cdot 3H_2O$	30	296	[13]	Capric Acid-Lauric Acid	19.2–20.3	144–150	[4]
$KF \cdot 4H_2O$	18.5	231	[13]	Capric Acid-Lauric Acid	19.09	141.5	[5]
$CaCl_2 \cdot H_2O$	29	190.8	[13]	Butyl stearate-palmitate	17–20	137.8	[4]
$Na_2SO_4 \cdot 10H_2O$	32	251	[13]	Palmitic Acid-Stearic Acid	32.1	151.6	[5]
$Mn(NO_3)_2 \cdot 6H_2O$	25.8	125.9	[13]	Capric Acid-Palmitic Acid-Stearic Acid	19.93	129.4	[4]
$Mn(NO_3)_2 \cdot 6H_2O$	25.5	148	[17]	Lauric Acid-Myristic Acid-Stearic acid	29.29	140.9	[5]
$K_2HPO_4 \cdot 4H_2O$	18.5	231	[17]	Capric Acid-1-dodecanol	27	126.9	[4]
$FeBr_3 \cdot 6H_2O$	21	105	[17]	$Ca(NO_3) \cdot 4H_2O$-$Mg(NO_3)_3 \cdot 6H_2O$	30	136	[17]
$LiNO_3 \cdot 2H_2O$	30	296	[17]	$CH3COONa \cdot 3H_2O$ + NH_2CONH_2	30	200.5	[17]
$LiBO_2 \cdot 8H_2O$	25.7	289	[17]	$CaCl_2$ + $NaCl$ +KCl+H_2O	26–28	188	[17]
$CaCl_2 \cdot 6H_2O$	29	191	[17]	$Na_2SO_4 \cdot 10H_2O$-$Na_2HPO_4 \cdot 12H_2O$	32.52	226.9	[5]

3.2. PCM Properties

For a material to be used as PCM, there are some thermal, physical, chemical, and kinetic properties that have to be satisfied. In addition, the final decision for a PCM to be manufactured for thermal storage also takes into account the environmental impact and the economics. Regarding the thermal properties, firstly, the phase transition temperature has to be in the desired operating temperature range in order to aver that the absorbing and releasing of heat happens in the desired range for the application temperature. Another very important characteristic for PCMs is high thermal conductivity of both liquid and solid phases, which is required in order to support the absorbing and releasing process of thermal energy. The latent heat of fusion per unit volume also has to be high, so that the storage density is higher and sensible storage can be achieved. In addition, high specific heat offers extra sensible heat storage. Finally, and also very important, is for the melting to happen congruently, in order for the phase change to be reproducible with stable cycling and long life [11,14,19].

On the physical properties needed, an important requirement is the minimum volume change during phase transition as well as small vapor pressure at the temperature of operation. Also, it is desired that the material has high density [14,19]. The kinetics of the charging/discharging cycle are also considerable because the insufficient nucleation rate leads to supercooling of the liquid phase, which is undesirable because the melting and solidification process take place in different temperatures. Therefore, high nucleation as well as enhanced crystal growth are required [8,14]. Apart from that, several chemical properties have to be satisfied by a PCM. Firstly, the freezing and melting cycle has to be completely reversible and the material should not degrade after a big number of cycles. Additionally, the material has to be noncorrosive to the environment used [11,14]. Finally, the materials used as PCMs have to be nontoxic, inflammable, and nonexplosive, both for safety and environmental reasons. From an economical point of view, they have to be available and abundant in nature, and cost-effective [19].

4. Improvement of Thermal Performance

4.1. Encapsulation

The practical use of solid-liquid PCMs impedes, due to some deterrents, characteristics, such as leakage of the liquid phase. To overcome these problems, new types of PCMs forms have been investigated, such as form-stable (or shape-stabilized) PCMs and encapsulated PCMs [20]. Encapsulation methods are used in order to overcome the leakage problems by keeping the material suitably contained inside a coating, a shell, or a matrix (form-stable PCMs). Encapsulation of PCMs can bring great improvement in enhancing the stability and heat transfer efficiency of the PCM, as well as facilitating its utilization, storage, and transportation.

The obstruction of material interaction with the environment not only prevents the leaking of the material, but also protects it from interacting with the environment, which could provoke degradation or corrosion during the melting-freezing cycles [2,21]. Furthermore, encapsulation increases the surface area of the PCM, which enhances the heat transfer efficiency. Moreover, the capsule form gives the capability of controlling the volume changes during melting, as well as the possibility of increasing thermal conductivity by inserting additives such as nanoparticles or C-based materials. Another great capability of manufacturing composite PCMs is that they can be applied directly in applications without any extra devices. For example, the fabrication of a PCM can be in forms such as powders or pastes, and they can be used directly as additives in constructional materials [21,22].

The encapsulated PCMs are composed of the core material, which is the PCM, and the shell material, which can be inorganic or organic. Their shape can be spherical, tubular, oval, or irregular, depending on the materials and the fabrication process. Additionally, a core-shell composite can have single or several cores within the capsule and multiple shells. A classification of encapsulated PCMs is related to their size. The PCMs can be characterized as macroencapsulated when their size is bigger than 1000 μm, microencapsulated when their size is between 1 and 1000 μm, and, finally, nanoencapsulated when their size is less than 1 μm [21]. The most interesting encapsulated PCMs to investigate are microencapsulated and nanoencapsulated PCMs because of the higher surface area per volume unit, which seems to facilitate thermal transfer and lead to better thermal performance. Another reason is that small particle sizes are more durable to mechanical stress [22].

4.1.1. Microencapsulation

Microencapsulated PCMs emerge from a coating process of the pure PCM. Droplets or individual particles of the PCM are coated with another material to form capsules of 1–1000 μm size. The core material can be any solid-liquid PCM, and the shell (container) material can be an inorganic or a polymer material. The most studied inorganic shells are SiO_2 and TiO_2 [23], while the organic shells used are poly(methyl methacrylate) (PMMA) [24], polystyrene [25], melamine-formaldehyde [26], and other, more complicated, structures [27]. There are numerous microencapsulation methods that can be used (Figure 3). The selection of the suitable microencapsulation method depends on the physical and chemical properties of both core and shell materials [22] and on the chemical environment for which they are intended to be used, for example in building [28].

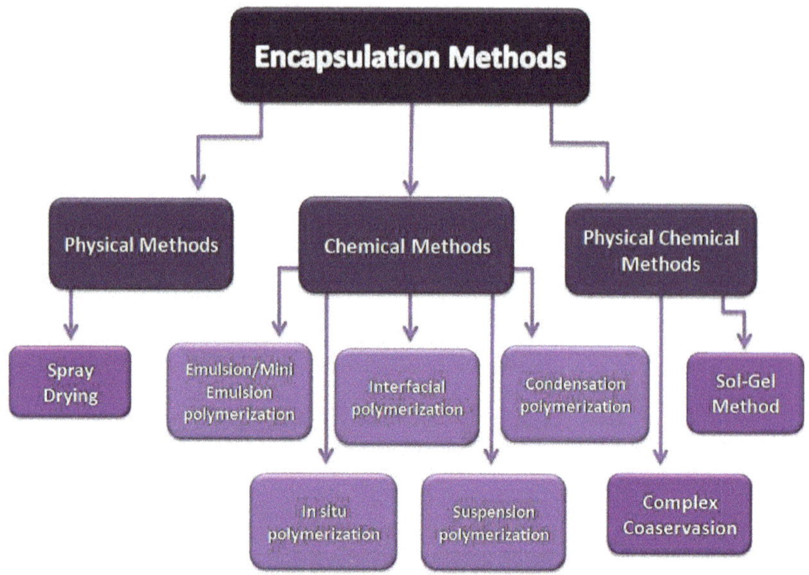

Figure 3. Physical, chemical, and physicochemical methods of encapsulation.

4.1.2. Methods of Microencapsulation

In physical methods, the material that creates the shell does not take part in any chemical reactions. The shell is mechanically applied on the core material by physical processes such as binding due to adhesion, drying, etc. There are a lot of physical methods, such as spray drying, vibrational nozzle, centrifugal extrusion, solvent evaporation, etc., although the method that is most used for microencapsulating PCMs is spray drying. The equipment and knowledge in this method is widely available. Additionally, the method can be easily scaled up, with average microcapsule size around some micrometers and efficient thermal performance; although, the high operating temperature, the agglomeration of microcapsules in the drying chamber, and the remaining uncoated particles are some its disadvantages [22,29].

In physicochemical synthesis methods, the capsule emerges through a physicochemical process. A physicochemical process is a combination of physical and chemical methods. The physical methods include processes such as cooling, heating, phase separation, etc., while the chemical methods include chemical processes such as condensation, hydrolysis, cross-linking, etc. Those methods combined result in microencapsulation. The most widely-used physicochemical methods are complex coacervation and sol-gel technique.

The complex coacervation method is suitable for fabrication of microcapsules with shell of organic materials such as chitosan, gum Arabic, gelatin, etc. This method is based on phase separation. The shell polymer is dissolved in an emulsion with the PCM (core material). The shell is formed around the droplet of the core material by evaporation of the solvent. In another way, the formation of the microcapsule can be managed by electrostatic coalescing of two different organic polymers around the droplet, and then decreasing the temperature and pH of the emulsion to initiate phase separation [30]. On the other hand, the sol-gel technique is used for the fabrication of inorganic shell microcapsules, such as SiO_2, TiO_2, etc. The sol-gel technique includes a sol solution, which is a solution of the hydrolyzed precursor compound, and an emulsion of the PCM. When the sol is added to the emulsion, the synthesis of the shell is happening on the PCMs droplet's surface, producing a gel of the microcapsules [31].

Finally, in chemical methods, the shell results from polymerization on the surface of a droplet or a particle of the core material. The most important methods are emulsion polymerization, suspension polymerization, interfacial polymerization, condensation polymerization, and in situ polymerization [22,31]. Suspension and emulsion polymerizations are widely used for PMMA and polystyrene shells, while the interfacial polymerization is preferred when the final shell consists of two or more monomers. In in situ polymerization, no monomers are used. The precursor compound hydrolyses in situ, and it initiates the polymerization. It is used for the synthesis of organic shell materials such as polyurea-formaldehyde and melamine-formaldehyde. Finally, condensation polymerization can also be used for inorganic shells, such as SiO_2, and it is usually part of other synthesis techniques, such as sol-gel [32].

4.1.3. Nanoencapsulation

Recent developments in nanotechnology, and specifically in synthesis methods of nanoparticles, allow the encapsulation of PCMs on nanoscale. Systems on nanoscale have essentially different physical and chemical properties to their counterparts on macroscale. The size effects of nanoscale provide outstanding properties to nanoencapsulated PCMs, which can make improvements to the efficiency of existing applications as well as meeting the requirements for new ones. For example, nanoencapsulation is proved to dramatically reduce the supercooling effect. Furthermore, the nanoencapsulated PCMs, due to their size, are able to form stable dispersions in fluids, known as nanofluids, without rapidly increasing their viscosity, which opens the path for new applications. In addition, nanoencapsulated PCMs have proven to be more durable to mechanical stress than larger capsules, which in some applications is very important [22,32].

The nanocapsule diameter is expected to be less than 1 µm, while in microencapsulated PCMs it is most typically 5 to 400 µm. As a result, not all the fabrication methods of microencapsulation that are presented in the previous section can be adopted for PCM capsules on nanoscale. The methods that are suitable for producing nanocapsules are mainly chemical. Specifically, the efficient synthesis techniques are emulsion polymerization, miniemulsion polymerization, interfacial polymerization, in situ polymerization, and sol-gel technique [22,32].

4.2. Form-stable PCMs

Solid-liquid PCMs, as known, present some disadvantages related to the liquid phase, such as leakage, volume change, supercooling, etc. To overcome these obstacles, composite PCMs are being investigated. A modification approach for pure PCMs, in order to enhance their performance and stability, is the fabrication of form-stable PCMs. Form-stable, or shape-stabilized, PCMs consist of a 3D matrix, or a porous structure, where the pure PCMs are entrapped. The matrix or porous structure can be in macro, micro, or nano scale [2,33]. The advantages of this technology are the limitation of liquid phase leakage, the enhancement of the thermal conductivity, the controlling of volume change during melting/freezing cycles, the reducing of PCM interaction with the environment, the avoidance of additional device requirement, and the ability of manufacturing the thermal energy system in the desired forms and dimensions for easy application. The materials used as substrates can be polymers or inorganic porous minerals [5,34].

4.2.1. Polymer Form-stable PCMs

In polymer-based form-stable PCM composites, the PCM is entrapped inside a cross-linked polymer matrix. The synthesis of the polymer matrix can be achieved with many different methods, providing structural diversity and, as a result, the ability of fabricating a product with special specifications, properly designed for each application. Additionally, polymeric cross-linked materials, as substrates, have shown great encapsulation rate results, which is a very important property [5].

There is a wide variety of the polymeric materials that can be used for this purpose. Materials that have been commonly used are polyethylene [35], polypropylene [36], polyurethane [37], poly methyl methacrylate [38], poly vinyl chloride [39], styrene-butadiene-styrene triblock copolymer [40], etc. Nevertheless, due to environmental concerns, there is great necessity for reducing the utilization of nonrenewable petroleum-based polymers. This has led to research for raw materials that can be received from natural sources, such as biomass, and they are renewable, biodegradable, environmentally friendly, of high abundance, and low cost. Such materials are celluloses, starches, and vegetable oils. Specifically, cellulose, agarose, chitosan [41], castor oil [42], etc. have been used for the fabrication of form-stable PCMs.

4.2.2. Inorganic Form-stable PCMs

Form-stable PCMs can also be fabricated with inorganic mineral porous materials such as kaolin, diatomite, vermiculite, sepiolite, bentonite, attapulgite, fly ash, opal, etc., as well as with expanded materials such as expanded graphite and expanded perlite. The porous structure of these materials makes them suitable for their utilization as carries for the encapsulation of the PCM. The fabrication of inorganic porous form-stable PCMs can be performed with impregnation method, melting intercalation technology, and melting adsorption method [43].

There is chemical compatibility between PCMs and the porous materials. The interactions developing between them are mainly surface tension, hydrogen bond, Van der Waals, etc. Due to these interactions, the form-stable composites show increased thermal stability and reliability. The absorbance rate, which is connected with latent heat capacity, is high in the case of expanded graphite, diatomite, and expanded perlite carriers. Thermal conductivity is vital for the energy storage and release efficiency. In mineral porous materials, thermal conductivity is relatively low but it can be easily enhanced with the addition of expanded graphite as an additive [43,44].

4.3. Thermal Conductivity Enhancement Techniques

Thermal conductivity in a thermal energy storage system is a very important factor for the efficiency of energy storing and releasing. PCMs', especially organics', main disadvantage is their low thermal conductivity. Low thermal conductivity decreases the heat transfer rate which means that the stored energy cannot be used efficiently, putting limitations on plenty of applications. As a result, the enhancement of thermal conductivity of PCMs is tremendously important for the amplification of system's performance [45]. The research that has been held for the improvement of thermal conductivity is mainly based on the improvement that encapsulation offers [10,46,47], the addition of highly-thermal conductive materials, such as carbon-based nanomaterials [48,49], metallic [50], or inorganic nanoparticles [51], and in the fabrication of form-stable PCMs with highly-thermal conductive carrier materials such as metallic foams [52], expanded graphite [53], etc. Another approach is the fabrication of nanofluid PCMs, which are dispersions of thermal conductive nanoparticles, for example, TiO_2, in the PCM liquid [54] (Table 5).

Table 5. Recent applications of micro- and nanoencapsulated PCMs with improved thermal conductivity.

Core Material	Shell Material	Encapsulation Method	Capsule Size	Thermal Conductivity of Pure PCM (W/(m.K))	Thermal Conductivity of Encapsulated PCM (W/(m.K))	Year	Ref
n-Eicosane	TiO_2	Interfacial polycondensation	Tubular: 1–5 μm length, 50–300 nm diameter Octahedral: 2–4 μm Spherical: 0.2–4 μm	0.161	1.244 (tubular) 1.023 (octahedral) 0.724 (spherical)	2019	[23]
N-octadecane	Poly(styrene-co-divinylbenzene-co-acrylamide)	Miniemulsion polymerization	-	-	-	2019	[27]
Myristic acid-Palmitic acid Eutectic	PMMA	Emulsion polymerization	0.1–70 μm	-	-	2019	[24]
Stearic Acid	$Si0_2$/GO	Sol-gel	2 μm	0.16	0.24 ($Si0_2$) 0.28 ($Si0_2$/GO)	2018	[49]
n-Dodecane Nanoalumina-Reinforced n-Dodecane	CNTs reinforced Melamine−Formaldehyde resin	In situ polymerization	-	0.14	0.297 0.219	2020	[55]
n-Octadecane	Boron Nitride reinforced Melamine-Formaldehyde	In situ polymerization	5–10 μm	0.14	0.11	2020	[56]
Paraffin	Graphene Oxide /Graphene nanoplatels	Self-assembly	-	0.25	0.90	2020	[57]
Stearyl Alcohol	SiO_2	Sol-gel	5–12 μm	0.14	0.15	2020	[58]
Methyl Laurate-based	Polyurethane	Pickering emulsion interfacial polymerization	8–10 μm	-	-	2020	[59]
n-Octadecane	SiO_2/graphene	Miniemulsion polymerization	256–473 nm	0.6416	1.4941	2019	[60]

Table 5. *Cont.*

Core Material	Shell Material	Encapsulation Method	Capsule Size	Thermal Conductivity of Pure PCM (W/(m.K))	Thermal Conductivity of Encapsulated PCM (W/(m.K))	Year	Ref
➤ n-Octadecane ➤ n-Eicosane	Crosslinked Polystyrene	Miniemulsion polymerization	136 nm 134 nm	0.22 0.18	0.12 0.11	2020	[61]
n-Octadecane	SiO_2	Miniemulsion polymerization	335 nm	0.15	0.38	2018	[62]
n-Eicosane-Fe_3O_4	SiO_2/Cu	Pickering emulsion interfacial polymerization	428–631 nm	0.4716	1.3926	2020	[63]

4.3.1. Enhancement of Thermal Conductivity by Encapsulation

According to the literature, organic PCMs have low thermal conductivity, which does not allow their efficient use in several applications. The preparation of core-shell or form-stable composite organic PCMs increases the thermal conductivity, as well as the heat transfer rate, due to increment of the surface area. A very important factor for the improvement of thermal conductivity is the size of the microcapsules. The extreme reduction of capsule size increases the thermal resistance because of the high contact rate of the capsules [47]. Another notable factor is the shell-to-core ratio. A higher shell-to-core ratio causes more increment on thermal conductivity but reduces the rate of energy storage because of the smaller amount of PCM inside the capsule [46]. The optimum shell-to-core ratio considering thermal conductivity, thermal storage, and mechanical strength is around 1:3 [64].

Inorganic shells, such as SiO_2 and $CaCO_3$, present better improvement on thermal conductivity than organic shells, such as PMMA, though they have decreased mechanical properties. For that reason, there has been investigation for fabricating core-shell PCMs with double shell: polymer shell for mechanical strength and inorganic for thermal conductivity enhancement [46]. Additionally, another point of view is the forming of a metallic second shell, which can improve thermal conductivity along with mechanical strength and agglomeration problems. Recent studies of micro/nanoencapsulation and form-stable composites are tabulated in Tables 5 and 6, respectively.

Table 6. Recent applications of form-stable composites PCMs.

PCM	Carrier	Preparation Method	Thermal Conductivity of Pure PCM(W/(m.K))	Thermal Conductivity of Encapsulated PCM (W/(m.K))	Year	Refs
PEG 10000	Graphene Aerogel/ Melamine Foam	Vacuum-assisted Impregnation	0.32	1.32	2020	[63]
Dodecane	Expanded graphite	Vacuum Infiltration	0.14	2.2745	2019	[65]
Lauric Acid Stearic Acid Eutectic	Carbonized Corn cob	Vacuum Impregnation	0.228	0.441	2019	[66]
PEG 1000	Halloysite NanoTube reinforced with Ag nanoparticles	Vacuum Impregnation	0.293	0.902	2019	[67]
Castor Oil	Polyurethane-Acrylate Oligomer	In situ polymerization	-	-	2019	[42]
m-Erythritol	Sepiolite and Exfoliated Graphite nanoplatelets	Vacuum Infiltration	0.372	0.756	2020	[68]
PEG 6000	Epoxy Resin porous Al_2O_3 ceramic	High-temperature blending and curing	0.393	2.54	2020	[69]

Table 6. *Cont.*

PCM	Carrier	Preparation Method	Thermal Conductivity of Pure PCM(W/(m.K))	Thermal Conductivity of Encapsulated PCM (W/(m.K))	Year	Refs
Capric Acid-Palmitic Acid Eutectic eutectic	Silica Xerogel/ Exfoliated Graphite nanoplatelets	Sol-gel	0.22	0.70	2020	[70]
Lauric Acid-based	➢ Carbon black/calcium carbonate powder ➢ MWCNTs/ calcium carbonate powder	Dispersion of nanoincusions in sonication bath	0.042 0.033	0.268 0.024	2020	[71]
PEG 4000	Almond shell Biochar	Vacuum impregnation	0.251	0.402	2018	[72]

4.3.2. Enhancement of Thermal Conductivity with Nanoparticle Additives

The enhancement of thermal conductivity of PCMs can be efficiently achieved with the fabrication of composite PCMs, which consist of the PCM material and a dispersion of high-thermal-conductivity additives. The PCMs mostly enhanced are organic PCMs such as paraffin, fatty acids, fatty acid ester, alcohols, and their eutectic mixtures. The additives can be carbon-based nanostructures [73,74], metallic nanoparticles, or nonmetallic nanoparticles. Besides the nature of the additives materials, the size, the shape, and the aspect ratio are the main factors that define the thermal conductivity improvement.

Carbon-based materials, such as carbon fibers (CF), carbon nanotubes (CNTs), graphene, graphite, and their derivatives, are widely being investigated for thermal conductivity enhancement, with very promising results. Carbon-based materials can be used as additives directly in the organic PCM, or they can be added inside a composite PCM, such as a form-stable PCM or a microcapsule, for further thermal conductivity improvement. CNTs' disadvantage is agglomeration, and as a result, CNTs show limited dispersion ability inside the PCM. This problem can be overcome with treatment of the CNTs, such as oxidation or doping, which weakens the intramolecular forces. On the other hand, carbon fibers show better dispersion ability, and they can be added efficiently without further treatment. Addition of graphite particles can also achieve great enhancement, although graphene nanoparticles are the most effective due to low interfacial thermal resistance and large aspect ratio [45,74].

Metallic nanoparticles can also be used as additives for the enhancement of thermal conductivity. The metallic nanoparticles that have been used are silver, copper, nickel, and magnesium nanoparticles. All of these metallic additives enhance efficiency of the thermal conductivity, but their practical use is limited because of a number of drawbacks that they show, such as low dispersion due to high agglomeration, low thermal stability, and high density [45,50].

Nonmetallic nanoparticles that can be added for thermal conductivity improvement are alumina (Al_2O_3), magnetite (Fe_3O_4), titania (TiO_2), mesoporous SiO_2, aluminum nitride (AlN), boron nitride (BN), etc. Nanoalumina particles improve the thermal conductivity, and the heat transfer, and they reduce the supercooling effect, although nanoalumina, along with titania, suffers from low dispersion problems [51]. Samaneh Sami and Nasrin Etesami [75] controlled the aggregation of TiO_2 nanoparticles inside paraffin PCM by adding the surfactant reagent stearoyl lactylate.

4.3.3. Enhancement of Thermal Conductivity with Metallic Foams and Expanded Graphite

Another perspective for thermal conductivity enhancement is the fabrication of a composite material consisting of the PCM and metallic foam as a carrier. Metallic foams that have been used for this purpose are copper foam [76], nickel foam [77], and porous graphite foam [78]. The composites can be prepared with vacuum infiltration or vacuum impregnation. These structures have plenty of advantages, such as high thermal conductivity, high thermal and chemical stability, high porosity, low density, and high aspect ratio. Their low density and high aspect ratio are the properties that make metallic foams prevail over metallic nanoparticles. Taking into account that thermal conductivity depends on the arrangement of the two components in the composite, it is mentioned that thermal conductivity is higher when the porosity is low because there are less interfacial interactions, and, as a result, the thermal resistance is lower. Although the energy storage capacity is improved when the porosity is higher, it is highlighted that both of those effects have to be considered for the optimization of the final material. Porous graphite and copper foam composite PCMs exhibit higher thermal conductivity in comparison with nickel foam composites [45].

Expanded graphite (EG) is a structure that comes from the oxidation and desiccation of graphite that has all the desired properties to be used for enhancement of thermal conductivity: chemical stability, high thermal conductivity, low density, and aspect ratio. The fabrication of the form-stable composite PCM can be achieved by impregnation of the molten PCM in the EG. The thermal conductivity of the composite depends on mass fraction, packing density, aspect ratio, surface area, and thickness of EG. High mass fraction and packing density leads to higher thermal conductivity [45,79,80].

5. Applications of PCMs in Buildings

In the past few decades, PCM technologies and the feasibility of their incorporation for thermal energy storage in the building sector have been investigated thoroughly. Thermal energy systems using PCMs are a sustainable and environmentally friendly solution that can definitely diminish the required energy consumption for the provision of comfort conditions of the building envelope. The PCM applications for thermal energy storage in this sector are divided in two categories: active and passive systems [12,81].

Active application systems based on PCMs require mechanical equipment or a source of additional energy for their operation, for example electricity to pumps or fans [82]. These systems are more suitable for cases where there is need for more heat transfer performance or better control of the application [64]. Active thermal storage systems with PCMs include systems located inside the building envelope for heating, ventilation, and air conditioning and systems that are integrated in the building structure and operating through air or water distribution, as well as systems located outside the building envelope, such as storage containers for domestic hot water, etc. [81,83]. On the other hand, passive application systems aim to improve the thermal performance of the construction systems such as the walls, the floor, the ceiling, etc. [84]. The most noteworthy passive applications of TES systems in building are presented in this section.

5.1. PCM Passive Application Systems in Building

Passive application PCM systems exploit the naturally-available energy sources along with the architectural design of building components to minimize the energy requirements of building operations. The main characteristic of these systems is that there is no requirement for mechanical equipment or additional energy, due to the fact that the heat is charged or discharged only because of temperature fluctuations when the environment's temperature rises or falls beyond the PCM's phase change temperature (Figure 4). Passive application PCM systems in the building sector can be classified as systems that are integrated into the building materials, such as materials for the walls, the floor, the ceiling, etc., and as systems that are implemented as components to the building envelope, such as blinds, suspended ceilings, etc. [16].

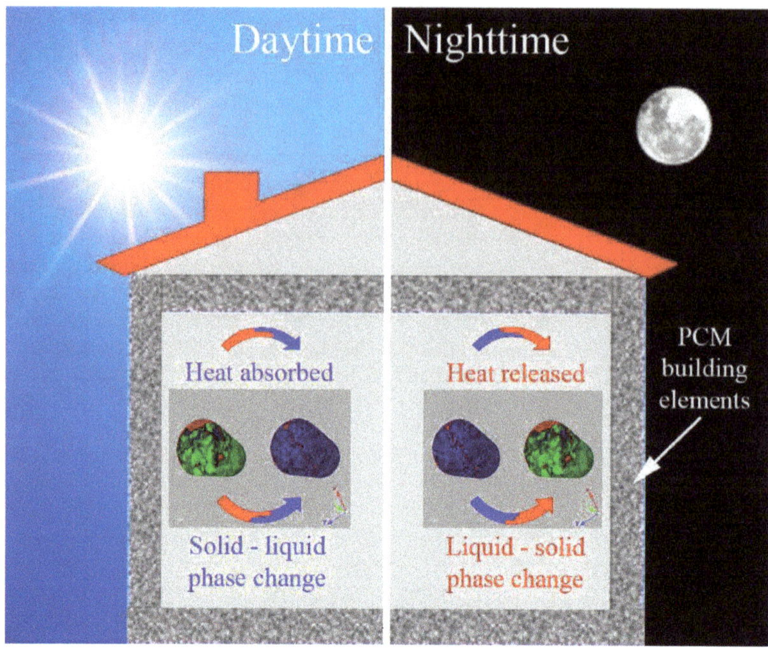

Figure 4. Charging process (working of PCM during daytime) and discharging process (working of PCM during nighttime).

5.1.1. PCMs in Wall

The most general and effective solution for building energy saving is the incorporation of PCM into constructional materials that are applied either to the internal or the external surface of the walls. PCMs can be integrated into mortars, gypsum boards, concrete blocks, bricks, and panels.

Mortars and Plasters

Mortars and plasters have been thoroughly investigated for PCM applications due to their use as coating materials for most of the building surfaces, along with their ability to have different compositions and their low cost. There are different kind of mortars, depended on the binder, such as gypsum, lime, and cement mortars [14]. The research that has been carried out about integration of PCMs in mortars and plasters is mostly about encapsulated PCMs, which involve high costs. As Cunha et al. [85] reported, nonencapsulated PCMs incorporated into mortar do not cause any significant enhancement to the mortar's properties. On the other hand, there are many studies showing that encapsulated PCMs into mortar and plaster have notable results for energy storage.

Sandra Cunha et al. [86] investigated the effects on mechanical, physical, and thermal properties of four different innovative mortars, by the direct integration of pure PCM. The physical and mechanical properties were affected, for example water absorption was reduced and mechanical strength was significantly changed above 10% of PCM. Thermal behavior amplifies in proportion with PCM amount. A total of 20% of PCM showed very efficient thermal performance and mechanical properties suitable for interior mortar application. Cynthia Guardia et al. [87] incorporated an encapsulated PCM to cement-lime mortar and studied both its thermal and mechanical properties. The addition of 10% and 20% of the PCMs remarkably increased the thermal properties for energy storage, while decreased the mechanical properties within the permissible limits. Hamed Abbasi Hattan et al. [88] studied the integration of polyethylene glycol (PEG) shape-stabilized in silica fume PCM in cement mortar for plastering building external brick walls. The experiment included thermal performance tests on outdoor temperature walls in two different cities. The thermal performance of the mortar was high, and the demands could be fulfilled without incorporating an amount of PCM that could cause decrease in mechanical strength, density, and water absorption.

Gypsum Boards

Another constructional element that can be used for TES are gypsum boards. The implementation of PCM technology into gypsum boards has been studied since 1990 because gypsum boards are low-cost and they are used widely in plenty of constructional applications. Srinivasaraonaik et al. [89] incorporated microencapsulated PCMs into gypsum mixture for gypsum boards. The addition of PCMs caused reduction in the compressive and flexural strength, while the porosity of the gypsum composites increased. As long as it concerns the thermal properties, thermal conductivity of the gypsum composites decreased in proportion with the amount of PCM incorporation. Similar results were obtained from the integration of graphene oxide microencapsulated PCM composites [73]. Additionally, in another study, Juan Pablo Bravo et al. [90] modified gypsum boards with a paraffinic microencapsulated PCM (20 wt%) and implemented a two-month enclosure test. The results were analyzed in a computer simulation program and they showed that microencapsulated PCMs with phase change temperature from 26.5 °C to 29.2 °C allow the charge and discharge of the material and, thus, the provision of indoor thermal comfort.

Concrete Blocks and Bricks

Another technology that has been investigated is the incorporation of PCMs into building elements such as concrete blocks and bricks. Vinh Duy Cao et al. [91] incorporated microencapsulated PCMs in Portland cement and in geopolymer concrete. The results obtained from thermal and mechanical analysis show that latent heat capacity increases, while thermal conductivity and compressive strength decreases, in proportion with the conciseness of PCMs. The addition of 3.2 wt.% and 2.7 wt.% to Portland cement and geopolymer concrete, respectively, can reduce the power consumption of the building in 23 °C by 11% and 15%, respectively. PCMs can also be successfully incorporated in bricks. Rajat Saxena et al. [92] integrated eicosane and OM35 PCMs into bricks. The results show an essential reduction of temperature fluctuation (4.5 °C to 7 °C) between incorporated and conventional bricks. A recent study from Mahmudul Hasan Mizan et al. [93] investigated the enhancement of interfacial bonding between concrete and PCM. The experiment included the preparation of 36 specimens with different surface preparation techniques each. The strength improvement was achieved by using silica fume.

PCM Panels and Wallboards

There are numerous applications of panels and wallboards of incorporated PCMs with conventional construction materials. In addition, there has also been research for the manufacturing of panels and wallboards of different materials, such as polymers [94,95] and metals [96]. Hang Yu et al. [97] investigated the fabrication of diatomite-based form-stable composite PCM wallboard for external building walls. Sayilacksha Gnanachelvam et al. [98] studied the utilization of tree wallboards (gypsum plasterboard, fiber cement board, and magnesium sulphate board) and a bio-PCM mat as fire resistant wallboards for their application to light-gauge steel-framed (LSF) wall systems made of cold-formed steel studs. Gypsum plasterboards provide better fire resistance that the other wallboards, and fiber cement boards provided the lowest. Behzad Maleki et al. [99] investigated the fabrication of plaster wallboards enhanced with nanocapsules with CuO nanoparticles. The PCM plaster wallboard reduced the thermal load of building by 13.83%.

5.1.2. Floor Applications

Building floors are another area where PCM technology can be applied for energy saving by thermal regulation. The applications that have been investigated for floor solutions are very varied. The addition of PCMs in the construction can be by adding a single layer of PCM [100,101] or by adding a multilayer of different types of PCMs [102,103]. Additionally, another approach is the development of floor panels with capillary network for PCM circulation [104], or the use of PCMs along with an electrical powered system for diminishing the demanded energy consumption [105]. Finally, PCMs can also be integrated in the raw materials used for construction, such as concrete [91]. Wanchun Sun et al. [84] investigated the thermal performance of a double-layer PCM radiant floor system containing two types of inorganic composite PCMs. The composites PCMs used were $Na_2HPO_4 \bullet 12H_2O@EG$ and $CaCl_2 \bullet 6H_2O@EG$. During winter, the thermal comfort lasted 2.2 times more than the reference room, which contained pebbles in the floor. During the summer, the thermal comfort lasted 1.7 times more. In both cases, the system can achieve shifting of the peak load. Shilei Lu et al. [86] studied the manufacturing of a PCM floor heating system consisting of a double pipe with three different modules. Regardless of the module, indoor thermal comfort was achieved. The average indoor temperature fluctuation was 1.8–3.0 °C.

5.1.3. Ceiling Applications

The development of ceiling solutions based on PCM technology has also been a field of research. There are many approaches for PCM ceiling applications, such as the incorporation of PCM ceiling panels into the roof construction [106,107], the development of ceiling panels with capillary network for PCM circulation [108], and the coupling of PCM technology with other technologies [109,110]. Mariana Velasco-Carrasco et al. [106] investigated the thermal performance of three different PCM blister panels for ceiling tiles application. The PCM used was INERTEK 23. The PCM was incorporated in three blister panels, two of which were enhanced with the addition of aluminum and steel wool particles, respectively. The panels were embedded between the roof slabs. The results obtained showed that the overall thermal performance of the aluminum- and steel-wool-particle-enhanced panels was improved, especially in the case of steel wool particles. Shilei Lu et al. [109] developed an innovative building cooling system of PCM ceiling coupled with earth-air heat exchanger (EAHE). The evaluation of experimental methods and data analysis indicated that indoor room temperature with this system could reduce peak temperature by 2.1 °C under 8-h timed cold storage experiment and 2.7 °C under 12-h timed cold storage experiment. Hansol Lim et al. [110] proposed a PCM integrated thermoelectric radiant cooling panel, which consists of thermoelectric modules, heat sinks, insulation, and PCM layer between two aluminum panels. The PCM layer provides passive cooling by freezing the PCM during the operation period for shifting the electrical load to the off-peak period demand periods.

5.1.4. Windows and Glazed Applications

Modern building designs contain large, glazed areas such as windows, glazed façade, and, sometimes, glazed roof, as they allow natural room lighting and vision, although glazing materials have very poor thermal performance which leads to increased energy consumption of the building. This indicates the necessity of expanding the PCM technologies for thermal energy storage to glazed materials also. Applications of PCM technology in glazed materials has been investigated. Changyu Liu et al. [82] studied the optical and thermal performance of a PCM-filled glazed unit in comparison with an air-filled glazed unit. The thermal performance of the PCM filled unit was highly improved, although the optical properties were degraded as the transparency was very low, especially when the PCM was at the liquid phase. An important factor to this matter is the thickness of the PCM in the glazed unit. Increased thickness improves the thermal performance but decreases the transparency. In this study it was found that the thickness should not be more than 16 mm in order to satisfy both thermal and optical demands. Corresponding study was held in reference [111–113].

6. Conclusions

A thermal energy storage system is a type of a sustainable energy storage system that is based on the utilization of materials that can store thermal energy when increasing their temperature and release it when the temperature is reduced. Latent heat storage systems using PCM are based on the absorption or release of heat that takes place during the material's phase transition (latent heat of fusion). There is a wide variety of compounds that are used as phase PCMs. For building applications, solid-liquid PCMs are preferred because of their ability to absorb and release large amounts of energy within a narrow range of temperature. Solid-liquid PCMs utilized in buildings can be classified into three major groups: organic compounds, inorganic compounds, and eutectics of organic and/or inorganic compounds. Organic PCMs gather the most advantageous properties.

The fabrication of composite PCMs enhances the thermal conductivity, thermal stability, and heat transfer rate, as well as impedes the leakage of the liquid phase. The encapsulated PCMs are composed of the core material (PCM) and the shell material, which can be inorganic or organic. A core-shell composite can have single or several cores within the capsule and multiple shells. The PCMs can be characterized by their size. The most interesting encapsulated PCMs are microencapsulated and nanoencapsulated PCMs, as the higher surface area per volume unit facilitates the thermal transfer and leads to better thermal performance. Encapsulation can be achieved with physical, chemical, or physico-chemical methods. The most widely-used methods for micro- and nanoencapsulation are chemical methods, such as emulsion, interfacial, and suspension polymerization, and by the physicochemical sol-gel technique.

Form-stable, or shape-stabilized, PCMs consist of a 3D matrix, or a porous structure, where the pure PCMs are entrapped. In polymer-based form-stable PCM composites, the PCM is entrapped inside a cross-linked polymer matrix. The synthesis of the polymer matrix can be achieved with many different methods, providing structural diversity and, as a result, the ability of fabricating a product with special specifications. Form-stable PCMs can also be fabricated with inorganic mineral porous materials, as well as with expanded graphite and perlite.

Thermal conductivity in a TES system is a very important factor for the efficiency of energy storing and releasing. As a result, the enhancement of thermal conductivity of PCMs is requisite. The improvement of thermal conductivity can be achieved with encapsulation, addition of highly-thermal-conductive materials such as carbon-based nanomaterials, metallic or inorganic nanoparticles, and the fabrication of form-stable PCM.

The application of TES systems in the building sector is a promising technology with many advantages. TES systems applied in buildings can be either active or passive. Active TES systems are characterized by forced convection heat transfer and, in some cases, also mass transfer. On the other hand, passive TES systems aim to create or maintain comfort conditions in the building, exploiting naturally-available energy sources, such as the solar energy or the wind, along with architectural design of building components. The main characteristic of these systems is that there is no requirement for mechanical equipment or additional energy due to the fact that the heat is charged or discharged only because of temperature fluctuations when the environment's temperature rises or falls beyond the PCM's phase change temperature.

Author Contributions: Conceptualization: C.V.P. and I.A.K.; methodology: C.V.P. and I.A.K.; software: C.V.P. and I.A.K.; validation: C.V.P. and I.A.K.; formal analysis: C.V.P. and I.A.K.; investigation: C.V.P. and I.A.K.; resources: C.V.P. and I.A.K.; data curation: I.A.K.; writing—original draft preparation: C.V.P. and I.A.K.; writing—review and editing: I.A.K. and C.A.C.; visualization: I.A.K. and C.A.C.; supervision: C.A.C.; project administration: C.A.C.; funding acquisition: C.A.C. All authors have read and agreed to the published version of the manuscript.

Funding: This research received no external funding.

Institutional Review Board Statement: Not applicable.

Informed Consent Statement: Not applicable.

Data Availability Statement: Data sharing not applicable.

Acknowledgments: Not applicable.

Conflicts of Interest: The authors declare no conflict of interest.

References

1. Sarbu, I.; Sebarchievici, C. A Comprehensive Review of Thermal Energy Storage. *Sustainability* **2018**, *10*, 191. [CrossRef]
2. Shchukina, E.M.; Graham, M.; Zheng, Z.; Shchukin, D.G. Nanoencapsulation of phase change materials for advanced thermal energy storage systems. *Chem. Soc. Rev.* **2018**, *47*, 4156–4175. [CrossRef]
3. Drissi, S.; Ling, T.-C.; Mo, K.H.; Eddhahak, A. A review of microencapsulated and composite phase change materials: Alteration of strength and thermal properties of cement-based materials. *Renew. Sustain. Energy Rev.* **2019**, *110*, 467–484. [CrossRef]
4. Cui, Y.; Xie, J.; Liu, J.; Wang, J.; Chen, S. A review on phase change material application in building. *Adv. Mech. Eng.* **2017**, *9*. [CrossRef]
5. Zhu, N.; Li, S.; Hu, P.; Wei, S.; Deng, R.; Lei, F. A review on applications of shape-stabilized phase change materials embedded in building enclosure in recent ten years. *Sustain. Cities Soc.* **2018**, *43*, 251–264. [CrossRef]
6. Nazir, H.; Batool, M.; Osorio, F.J.B.; Isaza-Ruiz, M.; Xu, X.; Vignarooban, K.; Phelan, P.; Kannan, A.M. Recent developments in phase change materials for energy storage applications: A review. *Int. J. Heat Mass Transf.* **2019**, *129*, 491–523. [CrossRef]
7. de Gracia, A.; Cabeza, L.F. Phase change materials and thermal energy storage for buildings. *Energy Build.* **2015**, *103*, 414–419. [CrossRef]
8. Sharma, A.; Tyagi, V.V.; Chen, C.R.; Buddhi, D. Review on thermal energy storage with phase change materials and applications. *Renew. Sustain. Energy Rev.* **2009**, *13*, 318–345. [CrossRef]
9. Keshteli, A.N.; Sheikholeslami, M. Nanoparticle enhanced PCM applications for intensification of thermal performance in building: A review. *J. Mol. Liq.* **2019**, *274*, 516–533. [CrossRef]
10. Drissi, S.; Ling, T.-C.; Mo, K.H. Thermal efficiency and durability performances of paraffinic phase change materials with enhanced thermal conductivity—A review. *Thermochim. Acta* **2019**, *673*, 198–210. [CrossRef]
11. Leong, K.Y.; Rahman, M.R.A.; Gurunathan, B.A. Nano-enhanced phase change materials: A review of thermo-physical properties, applications and challenges. *J. Energy Storage* **2019**, *21*, 18–31. [CrossRef]
12. Aziz, N.A.; Amin, N.A.M.; Majid, M.S.A.; Zaman, I. Thermal energy storage (TES) technology for active and passive cooling in buildings: A Review. *Matec Web Conf.* **2018**, *225*, 03022. [CrossRef]
13. Khadiran, T.; Hussein, M.Z.; Zainal, Z.; Rusli, R. Advanced energy storage materials for building applications and their thermal performance characterization: A review. *Renew. Sustain. Energy Rev.* **2016**, *57*, 916–928. [CrossRef]
14. da Cunha, S.R.L.; de Aguiar, J.L.B. Phase change materials and energy efficiency of buildings: A review of knowledge. *J. Energy Storage* **2020**, *27*, 101083. [CrossRef]
15. Kahwaji, S.; White, M.A. Edible Oils as Practical Phase Change Materials for Thermal Energy Storage. *Appl. Sci.* **2019**, *9*, 1627. [CrossRef]
16. Souayfane, F.; Fardoun, F.; Biwole, P.-H. Phase change materials (PCM) for cooling applications in buildings: A review. *Energy Build.* **2016**, *129*, 396–431. [CrossRef]
17. Heier, J.; Bales, C.; Martin, V. Combining thermal energy storage with buildings—A review. *Renew. Sustain. Energy Rev.* **2015**, *42*, 1305–1325. [CrossRef]
18. Ghadim, H.B.; Shahbaz, K.; Al-Shannaq, R.; Farid, M.M. Binary mixtures of fatty alcohols and fatty acid esters as novel solid-liquid phase change materials. *Int. J. Energy Res.* **2019**. [CrossRef]
19. Cabeza, L.F.; Castell, A.; Barreneche, C.; de Gracia, A.; Fernández, A.I. Materials used as PCM in thermal energy storage in buildings: A review. *Renew. Sustain. Energy Rev.* **2011**, *15*, 1675–1695. [CrossRef]
20. Rodríguez-Cumplido, F.; Pabón-Gelves, E.; Chejne-Jana, F. Recent developments in the synthesis of microencapsulated and nanoencapsulated phase change materials. *J. Energy Storage* **2019**, *24*, 100821. [CrossRef]
21. Milián, Y.E.; Gutiérrez, A.; Grágeda, M.; Ushak, S. A review on encapsulation techniques for inorganic phase change materials and the influence on their thermophysical properties. *Renew. Sustain. Energy Rev.* **2017**, *73*, 983–999. [CrossRef]
22. Jamekhorshid, A.; Sadrameli, S.M.; Farid, M. A review of microencapsulation methods of phase change materials (PCMs) as a thermal energy storage (TES) medium. *Renew. Sustain. Energy Rev.* **2014**, *31*, 531–542. [CrossRef]
23. Liu, H.; Wang, X.; Wu, D.; Ji, S. Morphology-controlled synthesis of microencapsulated phase change materials with TiO_2 shell for thermal energy harvesting and temperature regulation. *Energy* **2019**, *172*, 599–617. [CrossRef]
24. Sarı, A.; Bicer, A.; Alkan, C.; Özcan, A.N. Thermal energy storage characteristics of myristic acid-palmitic eutectic mixtures encapsulated in PMMA shell. *Sol. Energy Mater. Sol. Cells* **2019**, *193*, 1–6. [CrossRef]

25. Sánchez, L.; Sánchez, P.; de Lucas, A.; Carmona, M.; Rodríguez, J.F. Microencapsulation of PCMs with a polystyrene shell. *Colloid Polym. Sci.* **2007**, *285*, 1377–1385. [CrossRef]
26. Su, J.; Wang, L.; Ren, L. Fabrication and thermal properties of microPCMs: Used melamine-formaldehyde resin as shell material. *J. Appl. Polym. Sci.* **2006**, *101*, 1522–1528. [CrossRef]
27. Döğüşcü, D.K.; Damlıoğlu, Y.; Alkan, C. Poly(styrene-co-divinylbenzene-co-acrylamide)/n-octadecane microencapsulated phase change materials for thermal energy storage. *Sol. Energy Mater. Sol. Cells* **2019**, *198*, 5–10. [CrossRef]
28. Giro-Paloma, J.; Al-Shannaq, R.; Fernandez, A.I.; Farid, M.M. Preparation and Characterization of Microencapsulated Phase Change Materials for Use in Building Applications. *Materials* **2015**, *9*, 11. [CrossRef] [PubMed]
29. Alva, G.; Lin, Y.; Liu, L.; Fang, G. Synthesis, characterization and applications of microencapsulated phase change materials in thermal energy storage: A review. *Energy Build.* **2017**, *144*, 276–294. [CrossRef]
30. Onder, E.; Sarier, N.; Cimen, E. Encapsulation of phase change materials by complex coacervation to improve thermal performances of woven fabrics. *Thermochim. Acta* **2008**, *467*, 63–72. [CrossRef]
31. Su, W.; Darkwa, J.; Kokogiannakis, G. Review of solid-liquid phase change materials and their encapsulation technologies. *Renew. Sustain. Energy Rev.* **2015**, *48*, 373–391. [CrossRef]
32. Liu, C.; Rao, Z.; Zhao, J.; Huo, Y.; Li, Y. Review on nanoencapsulated phase change materials: Preparation, characterization and heat transfer enhancement. *Nano Energy* **2015**, *13*, 814–826. [CrossRef]
33. Kenisarin, M.M.; Kenisarina, K.M. Form-stable phase change materials for thermal energy storage. *Renew. Sustain. Energy Rev.* **2012**, *16*, 1999–2040. [CrossRef]
34. Alkan, C.; Sari, A. Fatty acid/poly(methyl methacrylate) (PMMA) blends as form-stable phase change materials for latent heat thermal energy storage. *Sol. Energy* **2008**, *82*, 118–124. [CrossRef]
35. Sobolciak, P.; Karkri, M.; Al-Maadeed, M.A.; Krupa, I. Thermal characterization of phase change materials based on linear low-density polyethylene, paraffin wax and expanded graphite. *Renew. Energy* **2016**, *88*, 372–382. [CrossRef]
36. Alkan, C.; Kaya, K.; Sari, A. Preparation, Thermal Properties and Thermal Reliability of Form-Stable Paraffin/Polypropylene Composite for Thermal Energy Storage. *J. Polym. Environ.* **2009**, *17*, 254–258. [CrossRef]
37. Tang, B.; Wang, L.; Xu, Y.; Xiu, J.; Zhang, S. Hexadecanol/phase change polyurethane composite as form-stable phase change material for thermal energy storage. *Sol. Energy Mater. Sol. Cells* **2016**, *144*, 1–6. [CrossRef]
38. Wang, L.; Meng, D. Fatty acid eutectic/polymethyl methacrylate composite as form-stable phase change material for thermal energy storage. *Appl. Energy* **2010**, *87*, 2660–2665. [CrossRef]
39. Jin, X.; Li, J.; Xue, P.; Jia, M. Preparation and characterization of PVC-based form-stable phase change materials. *Sol. Energy Mater. Sol. Cells* **2014**, *130*, 435–441. [CrossRef]
40. Chen, P.; Gao, X.; Wang, Y.; Xu, T.; Fang, Y.; Zhang, Z. Metal foam embedded in SEBS/paraffin/HDPE form-stable PCMs for thermal energy storage. *Sol. Energy Mater. Sol. Cells* **2016**, *149*, 60–65. [CrossRef]
41. Şentürk, S.B.; Kahraman, D.; Alkan, C.; Gökçe, İ. Biodegradable PEG/cellulose, PEG/agarose and PEG/chitosan blends as shape stabilized phase change materials for latent heat energy storage. *Carbohydr. Polym.* **2011**, *84*, 141–144. [CrossRef]
42. Wu, B.; Zhao, Y.; Liu, Q.; Zhou, C.; Zhang, X.; Lei, J. Form-stable phase change materials based on castor oil and palmitic acid for renewable thermal energy storage. *J. Therm. Anal. Calorim.* **2019**, *137*, 1225–1232. [CrossRef]
43. Lv, P.; Liu, C.; Rao, Z. Review on clay mineral-based form-stable phase change materials: Preparation, characterization and applications. *Renew. Sustain. Energy Rev.* **2017**, *68*, 707–726. [CrossRef]
44. Umair, M.M.; Zhang, Y.; Iqbal, K.; Zhang, S.; Tang, B. Novel strategies and supporting materials applied to shape-stabilize organic phase change materials for thermal energy storage—A review. *Appl. Energy* **2019**, *235*, 846–873. [CrossRef]
45. Qureshi, Z.A.; Ali, H.M.; Khushnood, S. Recent advances on thermal conductivity enhancement of phase change materials for energy storage system: A review. *Int. J. Heat Mass Transf.* **2018**, *127*, 838–856. [CrossRef]
46. Salunkhe, P.B.; Shembekar, P.S. A review on effect of phase change material encapsulation on the thermal performance of a system. *Renew. Sustain. Energy Rev.* **2012**, *16*, 5603–5616. [CrossRef]
47. Zhang, H.; Wang, X.; Wu, D. Silica encapsulation of n-octadecane via sol-gel process: A novel microencapsulated phase-change material with enhanced thermal conductivity and performance. *J. Colloid Interface Sci.* **2010**, *343*, 246–255. [CrossRef]
48. Harish, S.; Ishikawa, K.; Chiashi, S.; Shiomi, J.; Maruyama, S. Anomalous Thermal Conduction Characteristics of Phase Change Composites with Single-Walled Carbon Nanotube Inclusions. *J. Phys. Chem. C* **2013**, *117*, 15409–15413. [CrossRef]
49. Lin, Y.; Zhu, C.; Fang, G. Synthesis and properties of microencapsulated stearic acid/silica composites with graphene oxide for improving thermal conductivity as novel solar thermal storage materials. *Sol. Energy Mater. Sol. Cells* **2019**, *189*, 197–205. [CrossRef]
50. Lin, S.C.; Al-Kayiem, H.H. Evaluation of copper nanoparticles—Paraffin wax compositions for solar thermal energy storage. *Sol. Energy* **2016**, *132*, 267–278. [CrossRef]
51. Mohamed, N.H.; Soliman, F.S.; El Maghraby, H.; Moustfa, Y.M. Thermal conductivity enhancement of treated petroleum waxes, as phase change material, by α nano alumina: Energy storage. *Renew. Sustain. Energy Rev.* **2017**, *70*, 1052–1058. [CrossRef]
52. Huang, X.; Lin, Y.; Alva, G.; Fang, G. Thermal properties and thermal conductivity enhancement of composite phase change materials using myristyl alcohol/metal foam for solar thermal storage. *Sol. Energy Mater. Sol. Cells* **2017**, *170*, 68–76. [CrossRef]
53. Li, C.; Zhang, B.; Xie, B.; Zhao, X.; Chen, J.; Chen, Z.; Long, Y. Stearic acid/expanded graphite as a composite phase change thermal energy storage material for tankless solar water heater. *Sustain. Cities Soc.* **2019**, *44*, 458–464. [CrossRef]

54. Liu, Y.-D.; Zhou, Y.-G.; Tong, M.-W.; Zhou, X.-S. Experimental study of thermal conductivity and phase change performance of nanofluids PCMs. *Microfluid. Nanofluidics* **2009**, *7*, 579–584. [CrossRef]
55. Akeiber, H.; Nejat, P.; Majid, M.Z.A.; Wahid, M.A.; Jomehzadeh, F.; Famileh, I.Z.; Calautit, J.K.; Hughes, B.R.; Zaki, S.A. A review on phase change material (PCM) for sustainable passive cooling in building envelopes. *Renew. Sustain. Energy Rev.* **2016**, *60*, 1470–1497. [CrossRef]
56. Xia, Y.; Cui, W.; Ji, R.; Huang, C.; Huang, Y.; Zhang, H.; Xu, F.; Huang, P.; Li, B.; Sun, L. Design and synthesis of novel microencapsulated phase change materials with enhancement of thermal conductivity and thermal stability: Self-assembled boron nitride into shell materials. *Colloids Surf. A Physicochem. Eng. Asp.* **2020**, *586*, 124225. [CrossRef]
57. Zhou, Y.; Li, C.; Wu, H.; Guo, S. Construction of hybrid graphene oxide/graphene nanoplates shell in paraffin microencapsulated phase change materials to improve thermal conductivity for thermal energy storage. *Colloids Surf. A Physicochem. Eng. Asp.* **2020**, *597*, 124780. [CrossRef]
58. Zhu, C.; Lin, Y.; Fang, G. Preparation and thermal properties of microencapsulated stearyl alcohol with silicon dioxide shell as thermal energy storage materials. *Appl. Therm. Eng.* **2020**, *169*, 114943. [CrossRef]
59. Wang, Z.; Ma, W.; Hu, D.; Wu, L. Synthesis and characterization of microencapsulated methyl laurate with polyurethane shell materials via interfacial polymerization in Pickering emulsions. *Colloids Surf. A Physicochem. Eng. Asp.* **2020**, *600*, 124958. [CrossRef]
60. Zhu, Y.; Qin, Y.; Liang, S.; Chen, K.; Tian, C.; Wang, J.; Luo, X.; Zhang, L. Graphene/SiO$_2$/n-octadecane nanoencapsulated phase change material with flower like morphology, high thermal conductivity, and suppressed supercooling. *Appl. Energy* **2019**, *250*, 98–108. [CrossRef]
61. Zhao, M.; Zhang, X.; Kong, X. Preparation and characterization of a novel composite phase change material with double phase change points based on nanocapsules. *Renew. Energy* **2020**, *147*, 374–383. [CrossRef]
62. Zhu, Y.; Qin, Y.; Wei, C.; Liang, S.; Luo, X.; Wang, J.; Zhang, L. Nanoencapsulated phase change materials with polymer-SiO$_2$ hybrid shell materials: Compositions, morphologies, and properties. *Energy Convers. Manag.* **2018**, *164*, 83–92. [CrossRef]
63. Liao, H.; Chen, W.; Liu, Y.; Wang, Q. A phase change material encapsulated in a mechanically strong graphene aerogel with high thermal conductivity and excellent shape stability. *Compos. Sci. Technol.* **2020**, *189*, 108010. [CrossRef]
64. Su, J.; Ren, L.; Wang, L. Preparation and mechanical properties of thermal energy storage microcapsules. *Colloid Polym. Sci.* **2005**, *284*, 224–228. [CrossRef]
65. Song, Y.; Zhang, N.; Jing, Y.; Cao, X.; Yuan, Y.; Haghighat, F. Experimental and numerical investigation on dodecane/expanded graphite shape-stabilized phase change material for cold energy storage. *Energy* **2019**, *189*, 116175. [CrossRef]
66. Zhang, W.; Zhang, X.; Zhang, X.; Yin, Z.; Liu, Y.; Fang, M.; Wu, X.; Min, X.; Huang, Z. Lauric-stearic acid eutectic mixture/carbonized biomass waste corn cob composite phase change materials: Preparation and thermal characterization. *Thermochim. Acta* **2019**, *674*, 21–27. [CrossRef]
67. Song, S.; Qiu, F.; Zhu, W.; Guo, Y.; Zhang, Y.; Ju, Y.; Feng, R.; Liu, Y.; Chen, Z.; Zhou, J.; et al. Polyethylene glycol/halloysite@Ag nanocomposite PCM for thermal energy storage: Simultaneously high latent heat and enhanced thermal conductivity. *Sol. Energy Mater. Sol. Cells* **2019**, *193*, 237–245. [CrossRef]
68. Tan, N.; Xie, T.; Feng, Y.; Hu, P.; Li, Q.; Jiang, L.-M.; Zeng, W.-B.; Zeng, J.-L. Preparation and characterization of erythritol/sepiolite/exfoliated graphite nanoplatelets form-stable phase change material with high thermal conductivity and suppressed supercooling. *Sol. Energy Mater. Sol. Cells* **2020**, *217*, 110726. [CrossRef]
69. Wu, B.; Lao, D.; Fu, R.; Su, X.; Liu, H.; Jin, X. Novel PEG/EP form-stable phase change materials with high thermal conductivity enhanced by 3D ceramics network. *Ceram. Int.* **2020**, *46*, 25285–25292. [CrossRef]
70. Tan, N.; Xie, T.; Hu, P.; Feng, Y.; Li, Q.; Zhao, S.; Zhou, H.-N.; Zeng, W.-B.; Zeng, J.-L. Preparation and characterization of capric-palmitic acids eutectics/silica xerogel/exfoliated graphite nanoplatelets form-stable phase change materials. *J. Energy Storage* **2021**, *34*, 102016. [CrossRef]
71. Mishra, A.K.; Lahiri, B.B.; Philip, J. Carbon black nano particle loaded lauric acid-based form-stable phase change material with enhanced thermal conductivity and photo-thermal conversion for thermal energy storage. *Energy* **2020**, *191*, 116572. [CrossRef]
72. Chen, Y.; Cui, Z.; Ding, H.; Wan, Y.; Tang, Z.; Gao, J. Cost-Effective Biochar Produced from Agricultural Residues and Its Application for Preparation of High Performance Form-Stable Phase Change Material via Simple Method. *Int. J. Mol. Sci.* **2018**, *19*, 3055. [CrossRef]
73. Zhang, Y.; Wang, K.; Tao, W.; Li, D. Preparation of microencapsulated phase change materials used graphene oxide to improve thermal stability and its incorporation in gypsum materials. *Constr. Build. Mater.* **2019**, *224*, 48–56. [CrossRef]
74. Cheng, W.-L.; Li, W.-W.; Nian, Y.-L.; Xia, W.-d. Study of thermal conductive enhancement mechanism and selection criteria of carbon-additive for composite phase change materials. *Int. J. Heat Mass Transf.* **2018**, *116*, 507–511. [CrossRef]
75. Sami, S.; Etesami, N. Improving thermal characteristics and stability of phase change material containing TiO$_2$ nanoparticles after thermal cycles for energy storage. *Appl. Therm. Eng.* **2017**, *124*, 346–352. [CrossRef]
76. Zheng, H.; Wang, C.; Liu, Q.; Tian, Z.; Fan, X. Thermal performance of copper foam/paraffin composite phase change material. *Energy Convers. Manag.* **2018**, *157*, 372–381. [CrossRef]
77. Hussain, A.; Tso, C.Y.; Chao, C.Y.H. Experimental investigation of a passive thermal management system for high-powered lithium ion batteries using nickel foam-paraffin composite. *Energy* **2016**, *115*, 209–218. [CrossRef]

78. Sedeh, M.M.; Khodadadi, J.M. Thermal conductivity improvement of phase change materials/graphite foam composites. *Carbon* **2013**, *60*, 117–128. [CrossRef]
79. Sarı, A.; Karaipekli, A. Preparation, thermal properties and thermal reliability of palmitic acid/expanded graphite composite as form-stable PCM for thermal energy storage. *Sol. Energy Mater. Sol. Cells* **2009**, *93*, 571–576. [CrossRef]
80. Tang, Y.; Lin, Y.; Jia, Y.; Fang, G. Improved thermal properties of stearyl alcohol/high density polyethylene/expanded graphite composite phase change materials for building thermal energy storage. *Energy Build.* **2017**, *153*, 41–49. [CrossRef]
81. Gholamibozanjani, G.; Farid, M. A comparison between passive and active PCM systems applied to buildings. *Renew. Energy* **2020**, *162*, 112–123. [CrossRef]
82. Meng, E.; Cai, R.; Sun, Z.; Yang, J.; Wang, J. Experimental study of the passive and active performance of real-scale composite PCM room in winter. *Appl. Therm. Eng.* **2021**, *185*, 116418. [CrossRef]
83. Song, M.; Niu, F.; Mao, N.; Hu, Y.; Deng, S. Review on building energy performance improvement using phase change materials. *Energy Build.* **2018**, *158*, 776–793. [CrossRef]
84. Cunha, S.; Lima, M.; Aguiar, J.B. Influence of adding phase change materials on the physical and mechanical properties of cement mortars. *Constr. Build. Mater.* **2016**, *127*, 1–10. [CrossRef]
85. Cunha, S.; Leite, P.; Aguiar, J. Characterization of innovative mortars with direct incorporation of phase change materials. *J. Energy Storage* **2020**, *30*, 101439. [CrossRef]
86. Guardia, C.; Barluenga, G.; Palomar, I.; Diarce, G. Thermal enhanced cement-lime mortars with phase change materials (PCM), lightweight aggregate and cellulose fibers. *Constr. Build. Mater.* **2019**, *221*, 586–594. [CrossRef]
87. Hattan, H.A.; Madhkhan, M.; Marani, A. Thermal and mechanical properties of building external walls plastered with cement mortar incorporating shape-stabilized phase change materials (SSPCMs). *Constr. Build. Mater.* **2021**, *270*. [CrossRef]
88. Srinivasaraonaik, B.; Singh, L.P.; Sinha, S.; Tyagi, I.; Rawat, A. Studies on the mechanical properties and thermal behavior of microencapsulated eutectic mixture in gypsum composite board for thermal regulation in the buildings. *J. Build. Eng.* **2020**, *31*, 101400. [CrossRef]
89. Bravo, J.P.; Venegas, T.; Correa, E.; Álamos, A.; Sepúlveda, F.; Vasco, D.A.; Barreneche, C. Experimental and Computational Study of the Implementation of mPCM-Modified Gypsum Boards in a Test Enclosure. *Buildings* **2020**, *10*, 15. [CrossRef]
90. Cao, V.D.; Pilehvar, S.; Salas-Bringas, C.; Szczotok, A.M.; Rodriguez, J.F.; Carmona, M.; Al-Manasir, N.; Kjøniksen, A.-L. Microencapsulated phase change materials for enhancing the thermal performance of Portland cement concrete and geopolymer concrete for passive building applications. *Energy Convers. Manag.* **2017**, *133*, 56–66. [CrossRef]
91. Saxena, R.; Rakshit, D.; Kaushik, S.C. Phase change material (PCM) incorporated bricks for energy conservation in composite climate: A sustainable building solution. *Sol. Energy* **2019**, *183*, 276–284. [CrossRef]
92. Mizan, M.H.; Ueda, T.; Matsumoto, K. Enhancement of the concrete-PCM interfacial bonding strength using silica fume. *Constr. Build. Mater.* **2020**, *259*, 119774. [CrossRef]
93. El Omari, K.; Le Guer, Y.; Bruel, P. Analysis of micro-dispersed PCM-composite boards behavior in a building's wall for different seasons. *J. Build. Eng.* **2016**, *7*, 361–371. [CrossRef]
94. Zhu, N.; Liu, F.; Liu, P.; Hu, P.; Wu, M. Energy saving potential of a novel phase change material wallboard in typical climate regions of China. *Energy Build.* **2016**, *128*, 360–369. [CrossRef]
95. Meng, E.; Yu, H.; Zhou, B. Study of the thermal behavior of the composite phase change material (PCM) room in summer and winter. *Appl. Therm. Eng.* **2017**, *126*, 212–225. [CrossRef]
96. Yu, H.; Li, C.; Zhang, K.; Tang, Y.; Song, Y.; Wang, M. Preparation and thermophysical performance of diatomite-based composite PCM wallboard for thermal energy storage in buildings. *J. Build. Eng.* **2020**, *32*, 101753. [CrossRef]
97. Gnanachelvam, S.; Ariyanayagam, A.; Mahendran, M. Fire resistance of LSF wall systems lined with different wallboards including bio-PCM mat. *J. Build. Eng.* **2020**, *32*, 101628. [CrossRef]
98. Maleki, B.; Khadang, A.; Maddah, H.; Alizadeh, M.; Kazemian, A.; Ali, H.M. Development and thermal performance of nanoencapsulated PCM/ plaster wallboard for thermal energy storage in buildings. *J. Build. Eng.* **2020**, *32*, 101727. [CrossRef]
99. Entrop, A.G.; Brouwers, H.J.H.; Reinders, A.H.M.E. Experimental research on the use of micro-encapsulated Phase Change Materials to store solar energy in concrete floors and to save energy in Dutch houses. *Sol. Energy* **2011**, *85*, 1007–1020. [CrossRef]
100. Royon, L.; Karim, L.; Bontemps, A. Optimization of PCM embedded in a floor panel developed for thermal management of the lightweight envelope of buildings. *Energy Build.* **2014**, *82*, 385–390. [CrossRef]
101. Sun, W.; Zhang, Y.; Ling, Z.; Fang, X.; Zhang, Z. Experimental investigation on the thermal performance of double-layer PCM radiant floor system containing two types of inorganic composite PCMs. *Energy Build.* **2020**, *211*, 109806. [CrossRef]
102. Ansuini, R.; Larghetti, R.; Giretti, A.; Lemma, M. Radiant floors integrated with PCM for indoor temperature control. *Energy Build.* **2011**, *43*, 3019–3026. [CrossRef]
103. Lu, S.; Xu, B.; Tang, X. Experimental study on double pipe PCM floor heating system under different operation strategies. *Renew. Energy* **2020**, *145*, 1280–1291. [CrossRef]
104. Lin, K.; Zhang, Y.; Xu, X.; Di, H.; Yang, R.; Qin, P. Experimental study of under-floor electric heating system with shape-stabilized PCM plates. *Energy Build.* **2005**, *37*, 215–220. [CrossRef]
105. Velasco-Carrasco, M.; Chen, Z.; Aguilar-Santana, J.L.; Riffat, S. Experimental Evaluation of Phase Change Material Blister Panels for Building Application. *Future Cities Environ.* **2020**, *6*. [CrossRef]

106. Pasupathy, A.; Athanasius, L.; Velraj, R.; Seeniraj, R.V. Experimental investigation and numerical simulation analysis on the thermal performance of a building roof incorporating phase change material (PCM) for thermal management. *Appl. Therm. Eng.* **2008**, *28*, 556–565. [CrossRef]
107. Griffiths, P.W.; Eames, P.C. Performance of chilled ceiling panels using phase change material slurries as the heat transport medium. *Appl. Therm. Eng.* **2007**, *27*, 1756–1760. [CrossRef]
108. Lu, S.; Liang, B.; Li, X.; Kong, X.; Jia, W.; Wang, L. Performance Analysis of PCM Ceiling Coupling with Earth-Air Heat Exchanger for Building Cooling. *Materials (Basel)* **2020**, *13*, 2890. [CrossRef]
109. Lim, H.; Kang, Y.-K.; Jeong, J.-W. Application of a phase change material to a thermoelectric ceiling radiant cooling panel as a heat storage layer. *J. Build. Eng.* **2020**, *32*, 101787. [CrossRef]
110. Liu, C.; Wu, Y.; Zhu, Y.; Li, D.; Ma, L. Experimental investigation of optical and thermal performance of a PCM-glazed unit for building applications. *Energy Build.* **2018**, *158*, 794–800. [CrossRef]
111. Ismail, K.A.R.; Henríquez, J.R. Thermally effective windows with moving phase change material curtains. *Appl. Therm. Eng.* **2001**, *21*, 1909–1923. [CrossRef]
112. Silva, T.; Vicente, R.; Amaral, C.; Figueiredo, A. Thermal performance of a window shutter containing PCM: Numerical validation and experimental analysis. *Appl. Energy* **2016**, *179*, 64–84. [CrossRef]
113. Fokaides, P.A.; Kylili, A.; Kalogirou, S.A. Phase change materials (PCMs) integrated into transparent building elements: A review. *Mater. Renew. Sustain. Energy* **2015**, *4*, 1–13. [CrossRef]

MDPI
St. Alban-Anlage 66
4052 Basel
Switzerland
Tel. +41 61 683 77 34
Fax +41 61 302 89 18
www.mdpi.com

Applied Sciences Editorial Office
E-mail: applsci@mdpi.com
www.mdpi.com/journal/applsci